T0290889

Scanning Transmission Electron Microscopy

Scanning Transmission Electron Microscopy

Advanced Characterization Methods for Materials Science Applications

Edited by

Alina Bruma

CRC Press is an imprint of the
Taylor & Francis Group, an **informa** business

First edition published 2021
by CRC Press
6000 Broken Sound Parkway NW, Suite 300, Boca Raton, FL 33487-2742
and by CRC Press
2 Park Square, Milton Park, Abingdon, Oxon, OX14 4RN

Library of Congress Cataloging-in-Publication Data

Names: Bruma, Alina, editor.
Title: Scanning transmission electron microscopy : advanced characterization methods for materials science applications / edited by Alina Bruma.
Description: First edition. | Boca Raton : CRC Press, 2021. | Includes bibliographical references and index. | Summary: "This book focuses on explaining and applying the principles of machine learning-based techniques and advanced image processing methods currently used in the electron microscopy community suitable for handling large electron microscopy data sets and extracting structure-property information for various materials. It explains and exemplifies how to use these methods in order to interpret STEM images and extract suitable information in order to reveal material properties at the nanoscale"– Provided by publisher.
Identifiers: LCCN 2020043166 (print) | LCCN 2020043167 (ebook) | ISBN 9780367197360 (hardback) | ISBN 9780429243011 (ebook)
Subjects: LCSH: Scanning transmission electron microscopy--Data processing.
Classification: LCC QH212.S34 S226 2021 (print) | LCC QH212.S34 (ebook) | DDC 570.28/25–dc23
LC record available at https://lccn.loc.gov/2020043166
LC ebook record available at https://lccn.loc.gov/2020043167

ISBN: 978-0-367-19736-0 (hbk)
ISBN: 978-0-429-24301-1 (ebk)

Typeset in Times LT Std
by KnowledgeWorks Global Ltd.

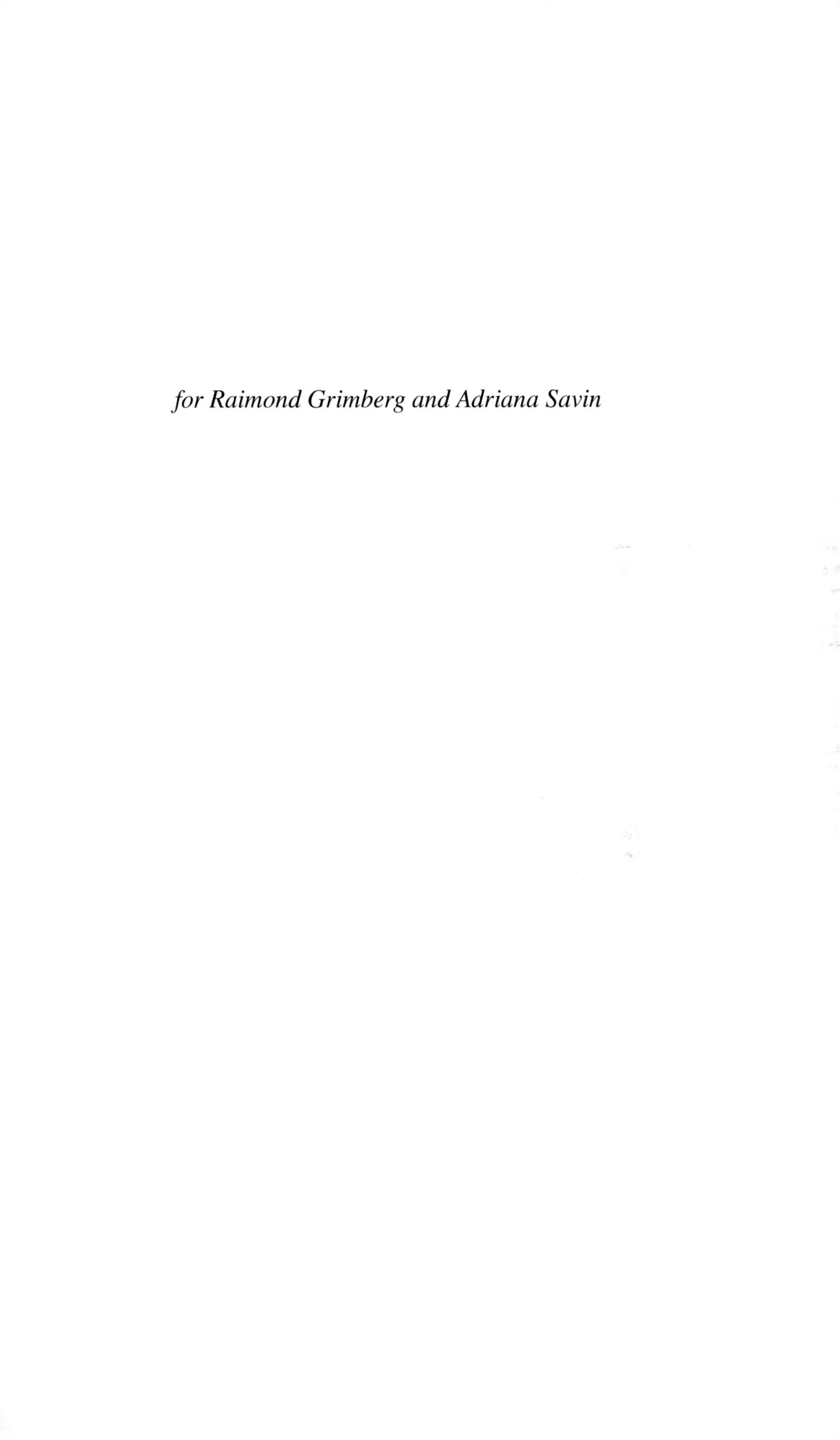

for Raimond Grimberg and Adriana Savin

Contents

Preface....................ix
About the Editor....................xi
Contributors....................xiii

Chapter 1 Practical Aspects of Quantitative and High-Fidelity STEM Data Recording....................1

Lewys Jones

Chapter 2 Machine Learning for Electron Microscopy....................41

Alex Belianinov

Chapter 3 Application of Advanced Aberration-Corrected Transmission Electron Microscopy to Material Science: Methods to Predict New Structures and Their Properties....................69

O. I. Lebedev

Chapter 4 Large Dataset Electron Diffraction Patterns for the Structural Analysis of Metallic Nanostructures....................111

Arturo Ponce, José Luis Reyes-Rodríguez, Eduardo Ortega, Prakash Parajuli, M. Mozammel Hoque, Azdiar A. Gazder

Index....................**147**

Preface

The fast progress in nanotechnology entails materials design down to nanoscale and requires characterization techniques powerful enough to help scientists understand their properties. Aberration-corrected scanning transmission electron microscopy (STEM) is, without a doubt, one of the most powerful characterization methods available to a materials' scientist. It allows the investigation of crystal structure at the local scale and complements in many ways powder diffraction techniques (e.g. x-ray or neutron diffraction), where data are acquired from large material volumes. However, unlike these techniques, a (S)TEM can access structural data in the real and reciprocal space. Recording and interpreting electron diffraction patterns holds the key to understanding the symmetry of the crystal, whereas adjacent techniques, such as imaging and spectroscopy, allow a direct visualization of the materials and provide access to their chemical composition. These adjacent spectroscopy techniques, namely energy-dispersive x-ray spectroscopy (EDXS) and electron energy-loss spectroscopy (EELS), provide unique information regarding a material's chemical fingerprint. Information such as chemical composition, oxidation state, coordination numbers of constituent atoms, and so forth are readily available to scientists, offering unprecedented access to information regarding materials.

Recently, significant advancements in instrumentation, hardware, and data processing have allowed many of the materials' properties to be mapped with ultra-high precision. Understanding these advancements is of the utmost importance for any materials scientist or microscopist. This book aims to do just that by bringing the latest developments in instrumentation, hardware, and data processing to you. The information comprised in these chapters is focused on discussing the latest approaches in the recording of high-fidelity quantitative annular dark-field (ADF) data, the unprecedented benefits brought by machine learning to electron microscopy, the latest advancements in imaging, and processing and data interpretation of materials notoriously difficult to analyze in (S)TEM. It also covers strategies to record, interpret, and understand large electron diffraction datasets for the analysis of metallic nanostructures.

We believe that this confluence of technologies is going to be a pivotal point in materials science research. We hope that the techniques presented by us will empower you to accelerate research in materials science and nanotechnology and motivate you to push the frontier of possibilities and architect a better future.

About the Editor

Dr. Alina Bruma received her Ph.D. degree in Nanoscale Physics from The University of Birmingham, United Kingdom in 2013. Dr. Bruma completed several postdoctoral stages at the Laboratory of Crystallography and Materials Science (CRISMAT-CNRS) in Caen, France; University of Texas at San Antonio, San Antonio, Texas; and The National Institute of Standards and Technology before moving to the American Institute of Physics Publishing in 2019. Her research has been focused on the study of crystalline structure of materials and the determination of their structure-property relationship using transmission electron microscopy and electron diffraction. Dr Bruma is also the Chairman of The Electron Diffraction Subcommittee at the International Center for Diffraction Data (ICDD).

Contributors

Dr Alex Belianinov is a Research and Development Staff Scientist at The Oak Ridge National Laboratory, Oak Ridge, Tennessee, where his work is focused on developing a research program centered around a variety of capabilities related to materials characterization, materials data analysis, and cleanroom chemistry. He has worked on developing analytical tools such as high-resolution ion microscopy, scanning probe microscopy, and electron microscopy to bridge the analysis gap between physical and data sciences. He is passionate about developing and utilizing data-processing algorithms to interface experiment and theory through a common high-performance computing hardware link. Dr. Belianinov has received several awards for his pioneering research including the R&D100 Award in 2016.

Dr. Lewys Jones is the Ussher Assistant Professor of Ultramicroscopy in the School of Physics, Trinity College Dublin, Ireland, and a Science Foundation Ireland & Royal Society University Research Fellow. He leads the Ultramicroscopy Research Group at the Advanced Microscopy Laboratory, focusing on instrument and technique development for high-performance electron microscopy, and is a codirector of the SFI-EPSRC Center for Doctoral Training in the Advanced Characterization of Materials. He received his Ph.D. which focused on scanning stability in the aberration-corrected scanning transmission electron microscope (AC-STEM) and on applications of focal-series annular dark-field data from the Department of Materials at the University of Oxford in 2013. He later developed two commercial software plug-ins for Digital Micrograph in collaboration with High Resolution Electron Microscopy (HREM) Research. Dr. Jones is a Member of the Institute of Physics and has been a Fellow of the Royal Microscopical Society since 2015.

Dr. Oleg Lebedev is a Research Director at the Laboratory of Crystallography and Materials Science (CRISMAT), The University of Caen, Lower Normandy (CNRS-ENSICAEN), France. Dr Lebedev received his Ph.D. in Physics and Mathematics at the Institute of Crystallography, The Russian Academy of Sciences in 1995. His current research interests are focused on the electron microscopy investigation of phase transition in nanomaterials, heteroepitaxial multilayer systems, atomic structure determination of interfaces, grain boundaries, and correlation with nanostructures' physical properties. Dr. Lebedev has authored more than 450 publications in peer review journals and has given more than 260 presentations at scientific meetings.

Dr. Arturo Ponce is an Associate Professor and Associate Chair in the Department of Physics and Astronomy at the University of Texas at San Antonio (UTSA). He leads the Structure Physics and Electron Microscopy Group at UTSA. He received his Ph.D. in Materials Science from the University of Cadiz, Spain, and completed a postdoctoral fellowship at Instituto de Física UNAM, Mexico, and CNRS, France. His research is focused on the study of the crystalline structure of materials and their physical properties through transmission electron microscopy and electron diffraction. It covers the theory and experimental aspects of the crystalline structure/physical property relationships based on the electron-matter phenomena.

1 Practical Aspects of Quantitative and High-Fidelity STEM Data Recording

Lewys Jones[1,2]
[1]Advanced Microscopy Laboratory, Centre for Research on Adaptive Nanostructures and Nanodevices (CRANN), Trinity College Dublin
[2]School of Physics, Trinity College Dublin

CONTENTS

1.1 Introduction 2
1.2 The Annular Dark-Field Detector 2
1.3 Example Applications of Quantitative ADF 3
1.4 Simulation Reference, Statistical Decomposition, or Hybrid Analysis Approaches 4
1.5 Practicalities of Recording ADF Detector Scans and Choosing Camera-Length 8
  1.5.1 Confocal or Swung-beam Detector Scanning Approaches 8
  1.5.2 Efficiency Response Across the ADF Detector Surface 10
  1.5.3 Detector Response Linearity 11
  1.5.4 Dropped Gain Versus Dropped Current 11
  1.5.5 Choosing Camera Length to Optimize ADF Inner- and Outer-Angle 12
1.6 Post-Specimen Flux Distributions and the Effect on Normalization Approach 13
1.7 Factors Affecting the Accuracy of Experimental Intensities and Peak Positions 16
  1.7.1 Sample Tilt 16
  1.7.2 Atomic-column "Cross Talk" 17
  1.7.3 Strain Contrast 18
  1.7.4 Amorphous Layers and Carbon Background 18
  1.7.5 Electron Dose and Pixel Size 19
  1.7.6 Cold Field-emission Current Fluctuation and Emission Decay 20
  1.7.7 Summary of Sources of Error 20

1.8 Environmental Noise and Scanning-Distortion in the Stem 20
1.9 Further Multi-Frame Applications of STEM Imaging, Spectroscopy, and 4D-STEM 24
1.9.1 Increasing Image SNR 24
1.9.2 Increasing Spectroscopic SNR 25
1.9.3 Increasing Pixel Density (Digital Super-resolution) 27
1.10 Future Possibilities in Quantitative ADF and Multiframe Imaging 28
1.10.1 New Geometries of Dark-field Detection 28
1.10.2 Single Electron Counting Binary ADF Imaging 30
1.10.3 Advances in Scan Patterning and Custom Scan-Design 30
1.11 Conclusion 31
Acknowledgments 31
References 32

1.1 INTRODUCTION

Aberration correction in the scanning transmission electron microscope (STEM) allows for the atomic resolution imaging of samples, while the incoherent nature of annular dark-field (ADF) allows for direct interpretation with a Z^n contrast relationship (Krivanek et al. 2010). More than only taking "pretty pictures", since the 1970s (Retsky 1974; Isaacson et al. 1976) there have been significant efforts to quantify this image intensity. At first, this growing field of "quantitative ADF" mostly referred to just the image intensity information; however, as it is often atomic-resolution data that are studied, spatial information such as peak-position shifts and strain mapping have recently started to be studied more. Spatial precision (and the effects of scan-distortion), while not the focus of this chapter, can affect image intensities, so it will be briefly discussed in the context of error analysis.

This chapter is structured as follows. first the hardware of the ADF detector is introduced along with some examples of how quantitative intensity imaging can be used. The mathematics of image normalization is given in the highlighted literature, so this review will instead concentrate on areas of best practice, experiment design, and error analysis in quantitative ADF. Next, the alternative approaches for data analysis are presented, namely either reference simulations, purely statistical approaches, or more recently a robust hybridization of the two. In this discussion of best-practice, Section 1.7 devotes significant discussion to the potential sources of error for practical microscopists. Finally, with quantitative imaging increasingly moving toward picometre-scale spatial precision measurements, Sections 1.8 and 1.9 introduce the use and application of multi-frame recording for the observation, diagnosis, and compensation of environmental noise and scan-distortion.

1.2 THE ANNULAR DARK-FIELD DETECTOR

The design of most ADF detectors is fairly similar. To facilitate retraction when not in use, these are usually mounted horizontally on a vacuum bellows. A scintillator inclined at 45° is attached to a light guide so that the signal reaches a horizontally

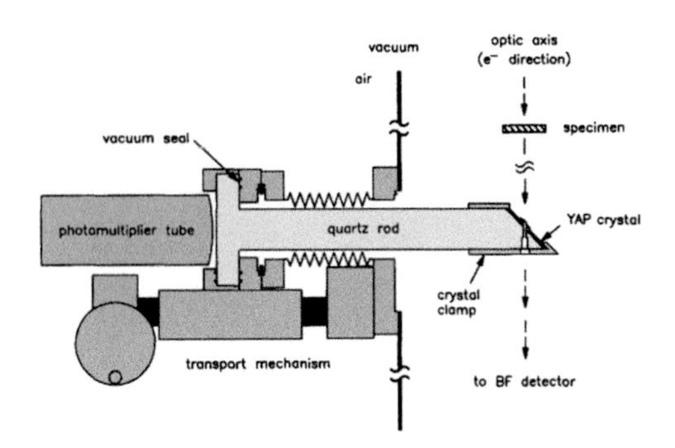

FIGURE 1.1 Left: Schematic of a typical ADF detector showing the hole in the scintillator to allow unscattered electrons through to other detectors or spectrometers. Right: Photograph of a Fischione 3000 ADF detector on a retractable mounting showing the metal-lined tube for unscattered electrons (image credit: (Jones 2016)).

mounted photomultiplier tube (PMT), Figure 1.1 (Kirkland and Thomas 1996). A hole through the scintillator and light guide (often lined with a metal tube to prevent charging) allows the un-scattered bright-field electrons to pass to another detector or spectrometer.

Scattered electrons hit the scintillator and produce optical photons which are detected and amplified at the PMT to produce the final signal. The output voltage from the photomultiplier has a dark-level, set by the amplifier brightness (or offset), which is added to the experimental signal; this combined signal is then amplified further by an amount depending on the amplifier's gain (or contrast) setting. Finally, this PMT voltage is averaged over the dwell-time specified by the user and passed through a so-called analogue-to-digital converter (ADC) (Grillo 2011). The output units are often displayed in "arbitrary counts"; importantly this should not be confused with real electron-counts. We will see later that while modern detector and amplifier systems are sufficiently sensitive to register single electron scattering impacts (Ishikawa et al. 2014), owing to practical limitations this is not presently implemented.

1.3 EXAMPLE APPLICATIONS OF QUANTITATIVE ADF

Extracting reliable information from quantitative ADF relies on being able to fix as many experimental factors as possible, ideally to leave only one free parameter to explore. Where thickness is constant (or presumed from extrapolation) quantitative ADF intensity can be used, with extensive simulation, to reveal local compositional variations, Figure 1.2 (Rosenauer et al. 2011). It should be noted that in this case the limiting error will be some combination of the reliability of thickness extrapolation and the effect of de-channeling or static-atomic-displacements.

However, as ADF imaging has a higher intrinsic scattering cross-section than electron energy-loss spectroscopy (EELS) or energy-dispersive x-ray spectroscopy (EDX) (around 100x and 10,000x more respectively), the signal-to-noise ratio (SNR) of quantitative ADF might yield a lower compositional error than these atomic resolution chemical mapping approaches directly, especially for samples that might

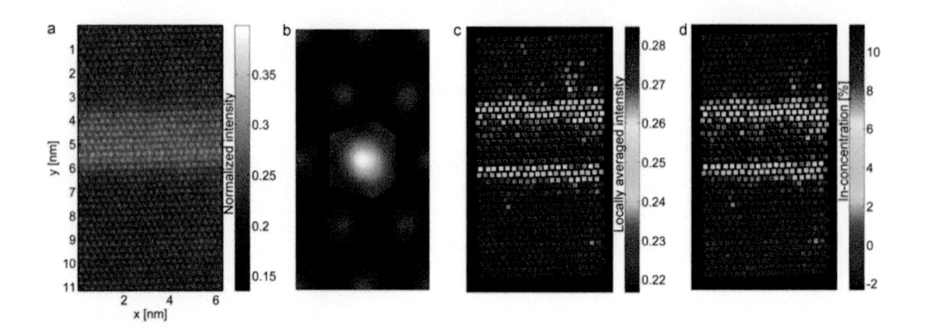

FIGURE 1.2 Part of an atomic-resolution experimental ADF image of InGaN (a) expressed in contrast units of "fractional beam". After peak-finding, integration polygons are defined (b), which are used to locally average the image intensity (c). Adjusting for thickness, and through comparison with extensive simulation, the resultant In-concentration map at atomic resolution is produced (d) (image credit: (Rosenauer et al. 2011)).

not tolerate the beam dose of spectroscopy. Where more than one imaging detector is used simultaneously, e.g. low-angle ADF (LAADF) with high-angle ADF (HAADF), and using calibration against other techniques such as x-ray diffraction (XRD), it is possible to map composition across far larger areas then EDX mapping alone (Ruiz-Marín et al. 2019).

For known (usually single element) composition samples, it is possible to determine variations in thickness or structure (LeBeau et al. 2010). When combined with energetic relaxation, these "atom counts" approaches can then yield reasonable estimates of three-dimensional (3D) particle morphology as an alternative to tomography (Figure 1.3) (De Backer et al. 2017).

Such inferred-tomography will never replace fully experimental multiviewing-direction tomography, but it can present a useful solution where samples cannot be multiply aligned either because of sample damage or holder-tilt limitations (Bals et al. 2012; Altantzis et al. 2019).

Whereas in tomography, recording additional images at different orientations may yield more spatial resolution, recording successive frames from the same viewing direction with quantitative ADF may be used to study dynamic processes that yield temporal resolution instead. This could be, for example, mass loss (Jones et al. 2014), surface reconstruction (Katz-Boon et al. 2015), crystal phase change (Pennycook et al. 2014), annealing (De Backer et al. 2017), or beam-induced motion (De wael et al. 2020).

In rare cases, where both thickness and composition can be presumed to remain near constant, the quantitative ADF signal can be used to infer the column occupancy of heavy species (Pennycook et al. 2014).

1.4 SIMULATION REFERENCE, STATISTICAL DECOMPOSITION, OR HYBRID ANALYSIS APPROACHES

Quantitative ADF is a rapidly evolving field being developed in several groups. First, let us briefly mention of some of the most common terms in use. The most crude unit of image intensity is the "arbitrary counts" output direct from the ADC. If the

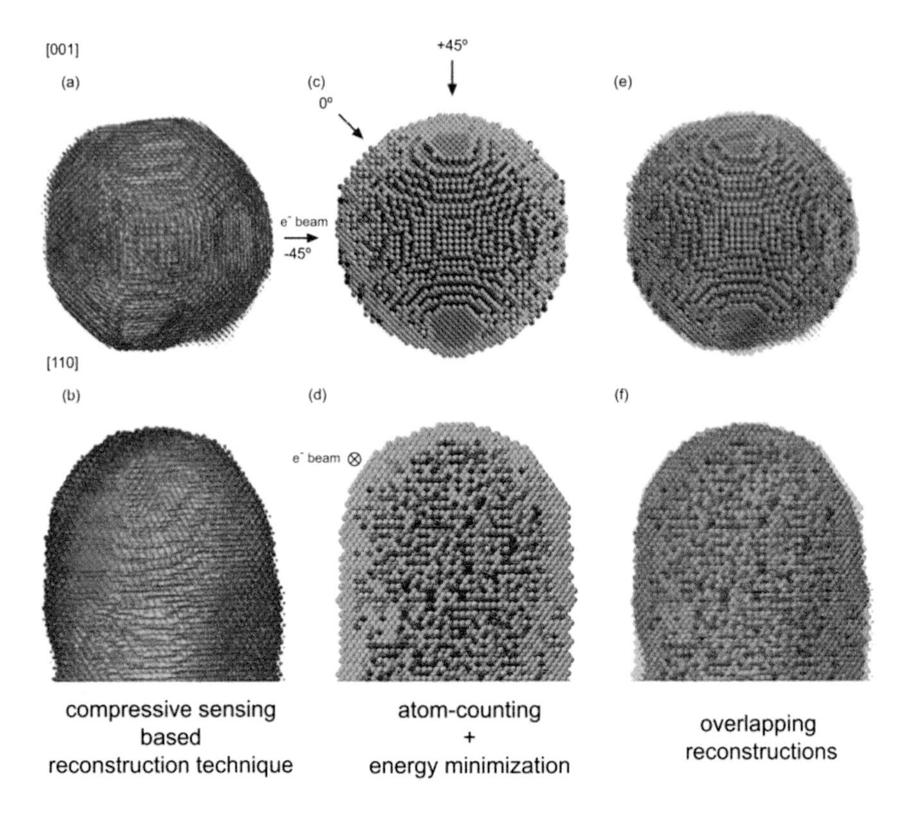

FIGURE 1.3 Comparison of two independent approaches for retrieving the 3D atomic structure of the tip of an Au nanorod. (a) The results of state-of-the-art compressed sensing tomography, (b) atom-counting followed by energetic relaxation, and (c) the comparison between the two approaches (image credit: (De Backer et al. 2017)).

detector linearity is acceptable (see later) and a suitable vacuum level is subtracted, then these units will at least be proportional to the scattered signal. Using such arbitrary-counts data, either "peak-intensities" can be measured or, because single-pixel measurements are often corrupted by noise, "averaged intensities" from within some small area around the peak can be taken.

Averaged intensities are more robust to noise, but the result depends on the size of the mask used for averaging. For peaked signals (resolved columns) a wider mask results in a lower average; alternatively, if units of "integrated intensity" are used up to a certain point a wider mask gives an *increased* value beyond which it plateaus (E et al. 2013). For isolated atoms, mask choice is flexible (Singhal, Yang and Gibson 1997; E et al. 2013), but in periodic structures this becomes more difficult and may be circles (LeBeau et al. 2010), squares (E et al. 2013), or Voronoi cells (Rosenauer et al. 2011; Jones et al. 2014; Nguyen, Findlay and Etheridge 2014). While the approach of Voronoi cells is relatively intuitive, and has the advantage that all image scattering is associated with some atomic column, for all but the thinnest samples, the image intensity rarely falls to zero between atomic columns, and simple Voronoi cells necessarily lead to some inaccuracy. The result of the analysis also depends on the pixel-sampling of the data (e.g. different results can be obtained if the same image is binned or upsampled), which indicates its potential weakness. One could imagine that for some other modalities, perhaps a watershed transformation would be appropriate to attribute electron scattering to each nearby atomic-column, but again this introduces a sharp arbitrary cut-off that is not physically representative of the scattering process.

To eliminate choosing any mask or imposing sharp cut-offs, fitted "gaussian volumes" can be used where peaks in the intensity data are fitted with Gaussian peaks (Van Aert et al. 2011; De Backer et al. 2016; Nord et al. 2017; Zhang et al. 2019). For thin samples that do not experience excessive beam broadening, Voronoi cells and fitted Gaussians are equivalent (De Backer, De wael, et al. 2015). While Voronoi cells are quick to calculate, in thicker samples or where large probe-tails exist, fitted Gaussians are more accurate.

If contrast ratios (Klenov and Stemmer 2006; Kotaka 2010) or statistical treatments are used as the analysis approach (Van Aert et al. 2011), no further preprocessing is required; however, for comparison with simulation the data must be normalized by the detector sensitivity.

Perhaps the most direct way to interpret experimental data is through comparison with image simulation. First, experimental data (in arbitrary counts) must be normalized by the mean detector efficiency, it can then be quoted in "fractional-intensities" (LeBeau et al. 2010) or "locally-averaged intensities" (Rosenauer et al. 2011). Reference images are simulated by carefully matching accelerating voltage, convergence angle, and detector angles for a range of sample conditions including thickness, composition, or both. Next, the analysis metric chosen from the list above is extracted from both the experimental data and the simulation library to allow an assignment on a column by column basis (Figure 1.4).

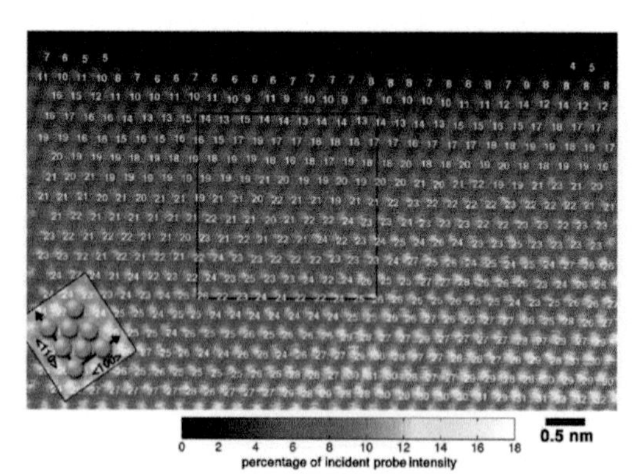

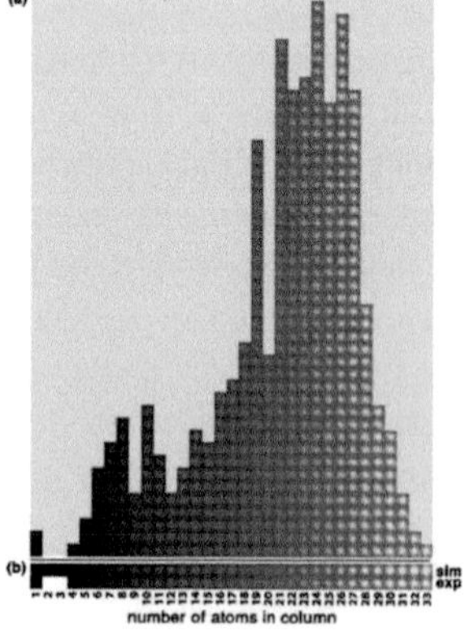

FIGURE 1.4 Left: Atomic-resolution ADF image of a [110] oriented Au wedge with contrast expressed in percentage of beam scattered. Right: The intensity at each atomic-column was averaged within a circular mask and compared with simulated values (image credit: (LeBeau et al. 2010)).

This direct method is intuitive but has one main drawback, systematic errors are difficult to detect. For example, some errors yield global scaling factors (see Section 1.7). This means that thickness or composition assignments made purely by comparison with simulated libraries can always find a match and the analysis reliability depends purely on the accuracy of the data normalisation. An additional complication is the wide range of simulation software packages to choose from, each of which gives very similar but sometimes slightly differing results for identical conditions (a comprehensive table listing many of these software can be found in Pryor, Ophus and Miao 2017).

Alternatively, and not necessarily requiring any detector normalisation, where a large number of nominally identical atomic columns are within the field of view, it may be possible to statistically decompose the intensities into their underlying grouping (sorted by atomic thickness) (Van Aert et al. 2011). This has been implemented in the StatSTEM software package (De Backer et al. 2016). Figure 1.5 shows the main stages in this process.

Finally, as image simulation may never capture all the subtleties of the experimental data, care must be taken in what metric to perform the matching on. Perhaps the "best practice" in quantitative ADF is the unit of the "scattering cross-section" (Retsky 1974; Isaacson et al. 1976; Rez 2001; E et al. 2013). This parameter is robust to magnification, defocus and convergence angle (E et al. 2013), source size (E et al. 2013), astigmatism and other aberrations (Martinez et al. 2014), scan noise (Jones and Nellist 2013), and small sample mis-tilt (MacArthur et al. 2015). The scattering cross-section then for an atom (or atomic column) is only dependent on the microscope's accelerating voltage and detector inner- and outer-angle.

The applicability of this approach is limited by two considerations: the need for a large number of observations per unique thickness, and for the spread of intensities

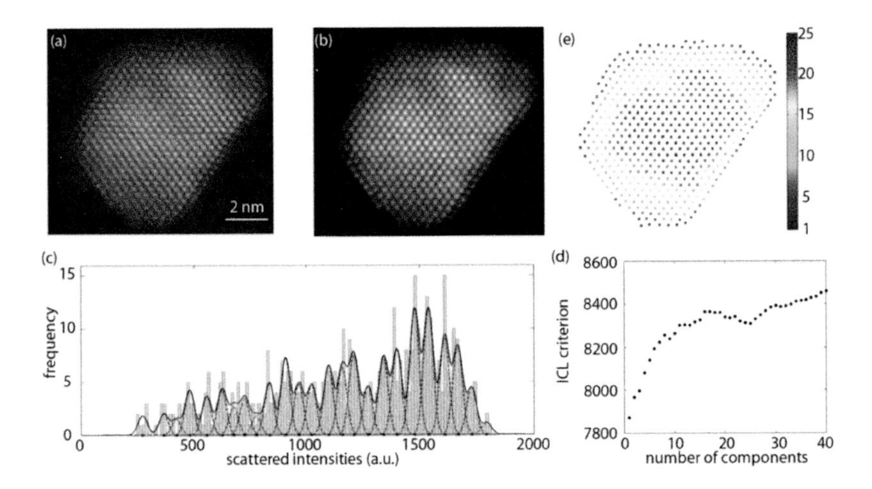

FIGURE 1.5 Stages in a statistical analysis of atomic resolution ADF data. (a) The experimental ADF image of a Pb nanoparticle embedded in an Si matrix. (b) The image data represented as a superposition of 2D gaussians after fitting. (c) The histogram of all gaussian volumes in the field of view. (d) The integrated classification likelihood (ICL) analysis of the intensities. (e) The atom-count assignments based on (c) and (d). Figure modified from (Van Aert et al. 2013).

from equal thicknesses to be significantly less than the difference between those of differing thicknesses (De Backer et al. 2013; De Backer et al. 2015). However, when these conditions are met, this approach is robust against systematic errors.

Now, what is rather elegant, is that the weaknesses of these two approaches are totally contrasting. Simulation comparison can analyze small fields of view, but it is not able to detect systematic bias; the statistical decomposition approach cannot analyze only small samples, but it is immune to normalization scaling errors. What is now considered the state-of-the-art approach is to combine these two techniques under one mathematical framework. This new hybrid approach delivers just that and is even able to help diagnose the often small (few percent) scaling errors that were previously undetectable by simulation reference alone (De wael et al. 2017).

1.5 PRACTICALITIES OF RECORDING ADF DETECTOR SCANS AND CHOOSING CAMERA-LENGTH

In Section 1.2 the construction of the ADF detector hardware was introduced. In this section, we will discuss how its response is measured and how its suitability for quantitative work is assessed. First, we describe the optical configurations available to map the ADF detector response, followed by an assessment of the detector's efficiency in collecting scattered electrons. Next, the importance of amplifier behavior (linearity) and the concept of "clipping" are discussed. Finally, once inner and outer angles have been verified (as opposed to relying on the manufacturer-supplied values), we look at choosing a camera length for optimal experiment design.

1.5.1 Confocal or Swung-beam Detector Scanning Approaches

There are fundamentally two approaches which can be followed when recording detector efficiency scans: a confocally scanned STEM probe or an angle-swung narrow parallel beam (sometimes called a pencil beam). The difference is discussed here, though both are recorded without a sample present.

In the confocal method the post-specimen optics are set to refocus the probe in the detector plane (FEI "image mode" or JEOL "STEM-align mode"), the scan-deflectors then scan the probe both in the sample plane and the detector plane (LeBeau and Stemmer 2008). Because of the post-specimen magnification achieved under these (TEM-like) conditions, confocal detector scans have dimensions of real-space millimetres. Once recorded an additional step is needed to scale the scan from millimetres to mill-radians (presuming a linear relationship) after comparing the shadow of the detector on the CCD or some other beam-deflection type approach (LeBeau and Stemmer 2008). While easier to record, and useful for quick assessments of the ADF, detector scanning in this way presents some issues. Firstly, as the beam passes through the objective lens (OL) it remains on-axis and does not replicate the high-angle scattering of ADF imaging. Secondly, it is dependent on the assumption that the millimeters to milli-radians conversion is linear out to the detectors edge (which has been shown from diffraction studies to be invalid beyond around 200 mrad (Craven et al. 2017)).

Figure 1.7 shows some examples of these "confocal" ADF scans, with the exception of one case (where scan-generator strength was insufficient) the whole detector surface is mapped and visible. The simplicity of this method may be why it is widely used in the normalized ADF papers already in the literature.

Although popular in the literature, the fact that the whole detector surface is visible in nearly all cases should perhaps be a cause for suspicion. This would imply that even on instruments with many post-specimen lenses (or even image-correctors) high angle rays up to and beyond 600 mrad all reach the detector chamber.

A passing mention in a figure caption of a far earlier work hints at an alternative option with a *tilted beam* (Singhal, Yang and Gibson 1997). When performing a tilted-beam scan, the condenser-lens system is adjusted to produce a cross-over in the back focal-plane of the OL and a narrow parallel pencil-beam at the specimen plane; the scan coils now swing this pencil beam in angle-space through the OL. A ray diagram comparing the two approaches is shown in Figure 1.6.

When using this method, the tilt-step of the scan can be precisely calibrated using a combination of the beams traced over the CCD (with ADF detector retracted) (Jones 2016) and the known diffraction angles from single-crystal (House et al. 2017) or polycrystalline (Jones et al. 2018b) reference materials. Once the tilt-step is calibrated, there are no extrapolation or scaling assumptions needed and the detector maps are natively returned with units of milli-radians. Furthermore, these scans now necessarily incorporate any effects of caustic aberrations (discussed later in Figure 1.10) or OL bore constriction, which dictates an upper bound for the detector

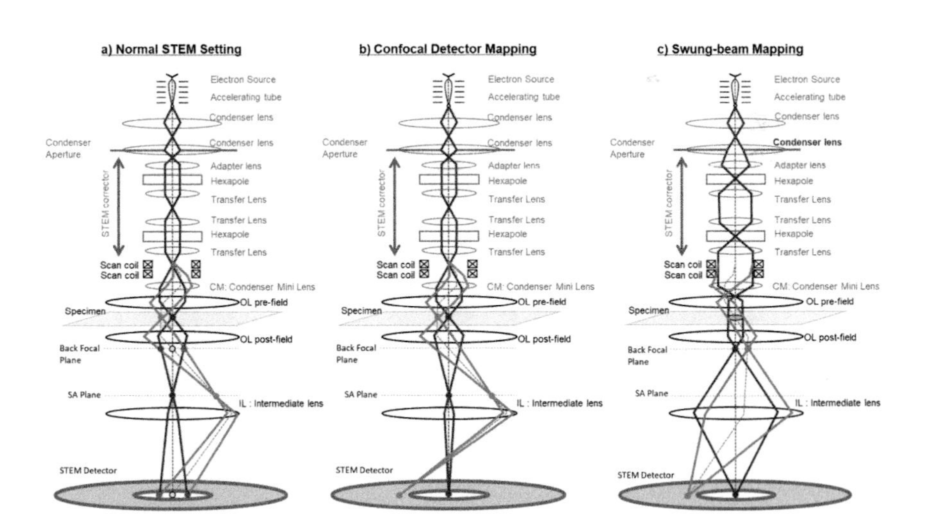

FIGURE 1.6 Optical ray diagrams to illustrate (a) regular STEM imaging, (b) the basic (confocal) detector scanning approach, and (c) the improved angle-space swung-beam mapping procedure. Note how when using confocal mapping the beam remains un-tilted thorough the objective, whereas with swung mapping the beam experiences the angular paths just as ADF scattered electrons would (including faithfully any aberrations or non-linear effects). Figure credit: Dr. H. Sawada and reproduced from (Jones et al. 2018b).

outer angle. These post-specimen aberrations are often ignored in STEM imaging (and certainly in image simulations); however, if we are to reliably quantify our experimental data these deserve further attention (Lotnyk et al. 2014; Martinez et al. 2015; Yamashita et al. 2015). An additional useful consequence of the swung-beam approach is that the angular size of EELS apertures (and their alignment behind the ADF detector) can be directly verified (Jones et al. 2018b).

To end this section, just one final cautionary note. As an unfortunate consequence of the somewhat "black box" nature of different vendors image-capture software, it is often advisable to record the detector sensitivity scans using the same dwell-time as the planned imaging experiment to avoid any possible effects of variable integration time (Ishikawa et al. 2014).

1.5.2 Efficiency Response Across the ADF Detector Surface

Ideally electron scattering to all angles (all points on the detector) would be recorded with equal sensitivity. Unfortunately, this is not always the case; the central hole in the scintillator means that light from beyond this reaches the PMT less efficiently. This generally results in a reduced efficiency at lower angles and must be accommodated via the comparison simulations (Rosenauer and Schowalter 2008), flux-weighting (Martinez et al. 2015), or by assuming a reduced effective outer-angle (Findlay and LeBeau 2013). Figure 1.7, shows a comparison of a range of common detectors currently used.

Of course the ideal solution is to improve the detector design, and some progress has already been made toward this by physically separating the light guide from the scintillator, elimination of the liner tube, or through the introduction of two-material detectors (Kaneko et al. 2014).

Manufacturer	Company 1			Company 2	Company 3	Company 4
	Detector A	Detector B	Detector C	Detector D	Detector E	Detector F
Detector Map	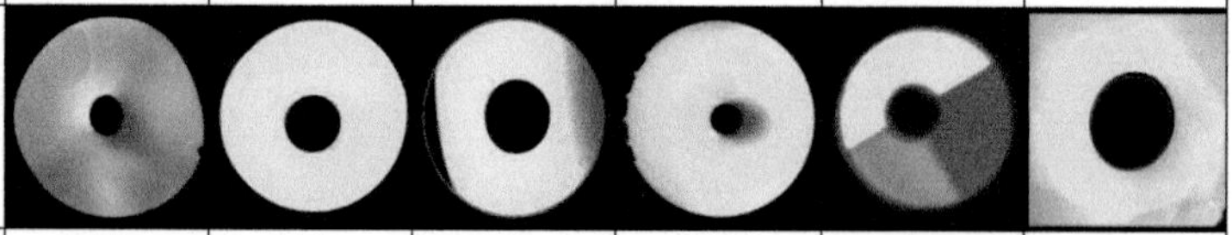					
Angle Ratio	5.47 x	3.09 x	2.89 x	5.91 x	2.90 x	3.50 x
'Flatness'	8.9 %	6.8 %	24.9 %	10.4 %	9.7 %	14.5 %
'Roundness'	8.2 %	5.4 %	10.6 %	5.4 %	28.1 %	2.4 %
'Smoothness'	30.0 % *	15.1 %	16.3 %	18.0 %	87.2 %	23.2 % *
'Ellipticity'	19.6 %	4.8 %	8.9 %	0.5 %	4.3 %	13.3 %
Average	16.7 %	8.0 %	15.2 %	8.6 %	32.3 %	13.4 %

FIGURE 1.7 Comparison of the sensitivity response of six commercially available ADF detectors, evaluated for various performance metrics. An ideal response would be a smooth concentric annulus of uniform response (figure credit and further interpretation (MacArthur, Jones and Nellist 2014)).

1.5.3 Detector Response Linearity

The term "detector linearity" is a somewhat ambiguous term used to describe several slightly different things. Fundamentally we need to evaluate the "detector-plus-amplifier" combination as a whole, and this depends greatly on the amplifier settings. Ideally the output should respond linearly with respect to brightness (LeBeau and Stemmer 2008), but also with respect to gain (Ishikawa et al. 2014; E et al. 2010), and perhaps most importantly, it must be linear with respect to beam-current (Yamashita et al. 2015) (Figure 1.8). Also remember, that as scattered electrons falling on the detector are Poisson distributed, not only does the mean intensity of thick regions need to be below the amplifier's saturation limit but also the high-intensity tail of this distribution (Grillo 2011).

Once the bounds of linearity are known, and working within these, the operator can set brightness towards the low end and contrast slightly higher (say 35% and 60% respectively for the below plots). This maximizes the available dynamic range without risking saturation in thick sample areas (which would lead to unwanted signal clipping).

As a result of the need to respect these linearities, the operator must choose an amplifier setup suitable for recording the detector scan in one of either two approaches, "dropped gain" or "dropped current" and these are discussed next.

1.5.4 Dropped Gain Versus Dropped Current

When imaging, total high-angle scattering may be around 10% of the primary beam and this is spread across the whole ADF detector. However, ADF normalization requires a detector sensitivity scan equivalent to 100% of beam current focused point by point across the detector. This leaves the operator with a dilemma and a choice, should the detector scan be recorded in so-called dropped gain or dropped current methods.

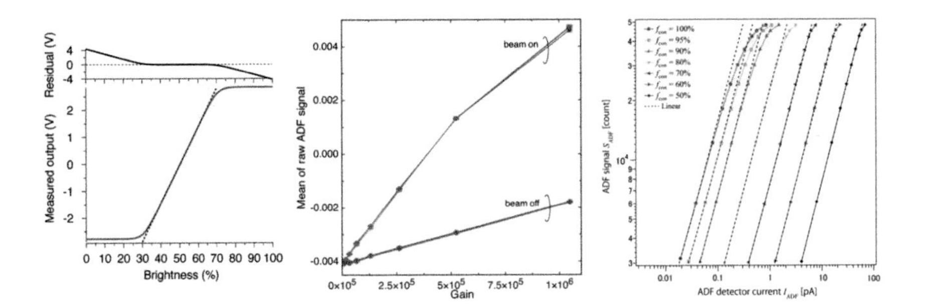

FIGURE 1.8 Evaluation of amplifier output linearity under various user-configurable conditions, including the brightness setting (left (LeBeau and Stemmer 2008)), the gain setting (middle (Ishikawa et al. 2014)), and with beam-current itself at various gain settings (right (Yamashita et al. 2015)). The operator should be aware of each of these reliable linear regimes for their own instrument.

In the dropped gain method, the full primary beam is used to map the ADF detector (LeBeau and Stemmer 2008), so to avoid massive amplifier saturation the gain settings must be set very low. To prevent errors arising from gain non-linearity, this same low gain must then be used for the imaging experiment, which may reduce sensitivity. There is also uncertainty about the behavior of the scintillator crystal (afterglow, depletion, saturation) with such differing illumination conditions between the imaging and mapping.

In the dropped current method, the amplifier is first optimized for the imaging experiment; then, by either choosing a smaller objective aperture or reducing the strength of the extraction anode, the beam-current is reduced by a factor of around 10. The detector is then mapped with this weaker beam, and the ratio of beam currents (as measured by either the CCD, or better, a Faraday cup/EELS drift-tube) incorporated in the normalization analysis (E et al. 2010).

Both are thought to yield roughly similar accuracies with the dominant error in the dropped gain method being the reliance on the linearity of output across such a wide range of electron dose, and in the dropped current method the precision with which the current ratio can be measured.

1.5.5 Choosing Camera Length to Optimize ADF Inner- and Outer-Angle

After considering the effect of objective-lens cut-off in the ADF detector mapping we now need to plan an experiment. Fortunately, the projector lens system in the STEM allows for a nearly free choice over camera-length and the inner-angle may be varied to optimize contrast between objects (Gnanasekaran, de With and Friedrich 2018). However, image simulations provide a more systematic guide for experiment design to maximize the precision of the recorded data (Figure 1.9). To compare ADF

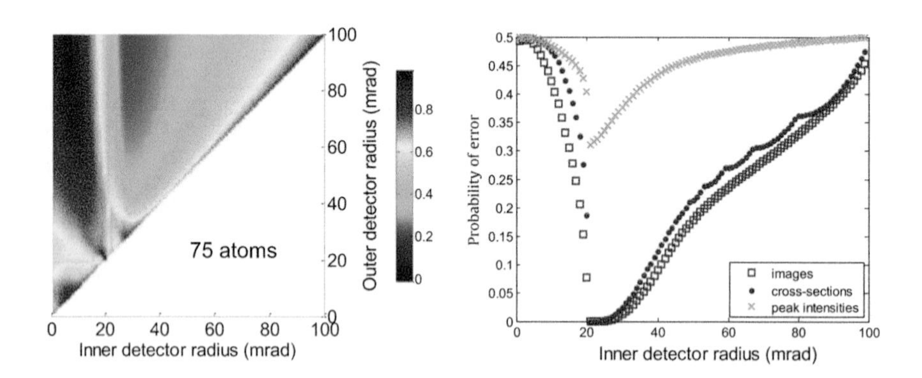

FIGURE 1.9 An example of a simulation-guided experiment design optimization study for determining the thickness of a 75-cell slab of $SrTiO_3$ by atom-counting. Left: The probability of error by miscounting ±1 atom where a free choice of inner- and outer-angle might be used. Right: a plot for a fixed 100 mrad outer-angle detector (for example, due to some fixed aperture or cut-off) where only inner angle can be varied. Note the probability of error for fitted image gaussians, cross-sections, and peak-intensities (figures credit: (De backer, De wael, et al. 2015)).

data against simulations, inner and outer detector angles must be precisely measured (LeBeau and Stemmer 2008; Martinez et al. 2015) for each of the required camera-lengths (Lotnyk et al. 2014).

For compositional selectivity at fixed thickness, the Z^n relationship of ADF scattering (Krivanek et al. 2010) improves with higher inner angles (Walther 2006; Wang et al. 2011). However, for light elements scattering to high angles is so weak the finite SNR becomes the dominant source of error, so for thickness measurements at fixed composition lower inner angles are preferable (Gonnissen et al. 2014; De Backer et al. 2015; Hovden and Muller 2012).

Once such an experiment-design optimization has been performed, the operator should choose an appropriate camera-length. In special cases, a continuously variable outer-angle setup can be achieved by using a special iris-aperture, but this is uncommon (Müller-Caspary et al. 2016).

1.6 POST-SPECIMEN FLUX DISTRIBUTIONS AND THE EFFECT ON NORMALIZATION APPROACH

Having discussed the asymmetry in the efficiency response of ADF detectors above, next we should consider the electron distribution that falls on them. Rutherford type scattering, which forms our dark-field images, is heavily forwardly biased (Forbes et al. 2011). We can visualize this in a qualitative way if the scattering through a sample is viewed on the CCD with the ADF retracted (Figure 1.10).

Figure 1.10 shows examples of scattering from a variety of relatively thick arbitrary samples for several common instrument types (Jones 2016). For un-corrected and probe-corrected instruments this flux-distribution is predominantly round; however, TEM-corrected instruments yield more complicated patterns owing to the additional nonround postspecimen lenses in the image corrector (Yamashita et al. 2015). With the addition of an in-column energy-filter this is further distorted in one direction. First, we can conclude from this these patterns that instruments with round optics only between the specimen and the detector are expected to give the most accurate quantitative ADF match to simulations (which themselves include no post-specimen optics). Next, in all cases an "outer cut-off" can be seen and, depending on the instrument, is between around 130 and 250 mrad indicating that below a certain camera-length the ADF outer-illumination angle will not increase any further (Lotnyk et al. 2014; Martinez et al. 2015; Yamashita et al. 2015). Many flux patterns also exhibit a brightened outer rim as the highest angle scattering is inverted back toward the center (Craven et al. 2017; Jones et al. 2018b). These factors all affect the appropriate outer-angle that must be used in image simulations. This is of course a related phenomenon to that seen with the angle-space swung mapping discussed earlier. As a result of these flux distributions and the already inhomogeneous response of the ADF detector (Figure 1.7), we must consider the interaction of these factors.

With only a very small percentage of high-angle scattering out beyond 200 mrad, the interaction of flux-asymmetry and detector inhomogeneity is most acute at the low angles and the inner edges of the ADF detector. Unfortunately, for many ADF detectors, this is precisely the region where their response is the most irregular (Figure 1.7).

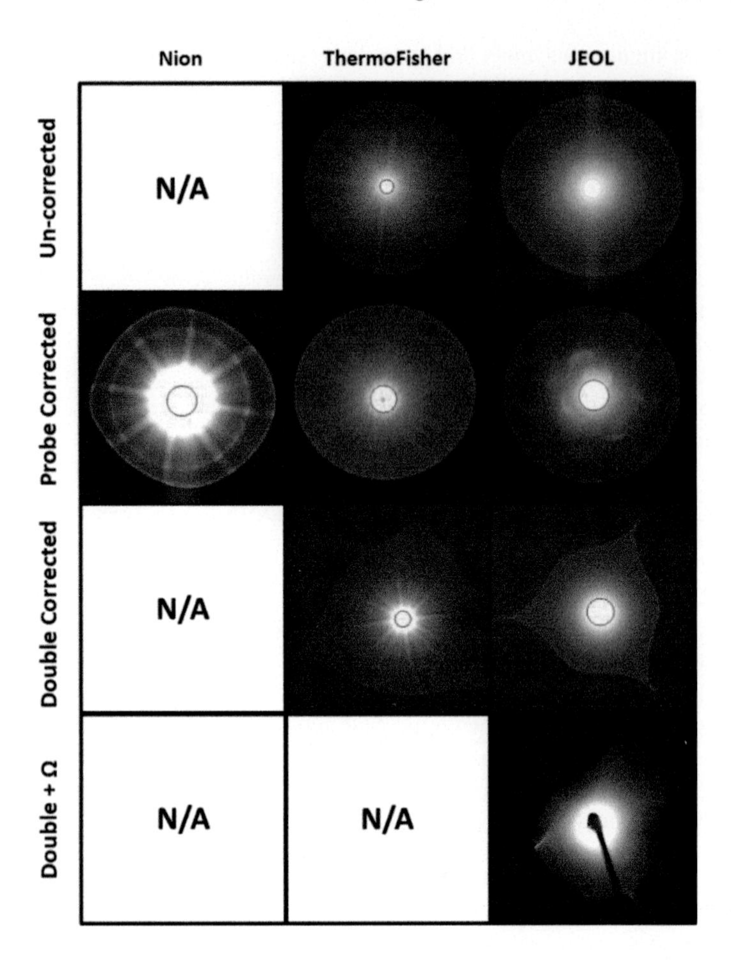

FIGURE 1.10 Various experimental STEM flux maps recorded at the CCD for several common instruments currently in use. In all cases the logarithm of the intensity is shown to improve the visibility of the dynamic range which spans around three orders of magnitude. Central solid circles indicate the BF disk (≈10 mrad for uncorrected, 20 mrad otherwise) (figure credit: Jones 2016).

This introduces another choice for the experimentalist when striving for the best possible match to simulations: should this hardware defect be accounted for in the reference simulations or in the experimental data normalization? A more in-depth discussion of this issue is given in Martinez et al. (2015), but an overview of the two approaches to solve this issue are as follows. If using the detector sensitivity profile (DSP) approach, then the whole active area of the detector is considered equally when determining its average signal response (averaged within some mask), and the experiment is normalized using this value. Reference simulations for this approach must then include a detector profile (typically one dimensional (1D)) specific to the individual detector of one particular instrument. However, only some simulation packages can accommodate such a DSP in the creation of the library data (Rosenauer and Schowalter 2008). While this approach allows for a simple experimental normalization to be used (LeBeau and Stemmer 2008), it precludes using other simulation codes that cannot introduce the DSP (Kirkland 1998; Ishizuka 2002; Koch 2002; Forbes et al. 2010), and cannot account for the fact that high-angle scattering is often truncated by post-specimen optics and requires a correction by imposing some artificial effective outer-angle (Findlay and LeBeau 2013). Alternatively, a two-dimensional (2D) sensitivity response of the detector (such as those shown in Figure 1.7) can be multiplied with an experimental flux distribution(e.g. Figure 1.10), complete with any post-specimen optical effects, to give an "electron flux-weighted" (EFW) detector efficiency. This approach recognizes the fact that the inner part of the detector collects more scattered electrons and contributes more signal to the final image (Jones et al. 2014; Martinez et al. 2015). With the experiment normalized in this way, the researcher is free to use any simulation software of their choosing and is able to compare scattering cross-sections between instruments with no further scaling required.

An example of these normalization masks and simulation parameters is shown in Figure 1.11 (Martinez et al. 2015).

Of course, the two approaches should not be mixed; for example, the sensitivity inhomogeneity should not be accommodated by *both* flux-weighting the experimental normalization and the reference simulation as this will lead to errors.

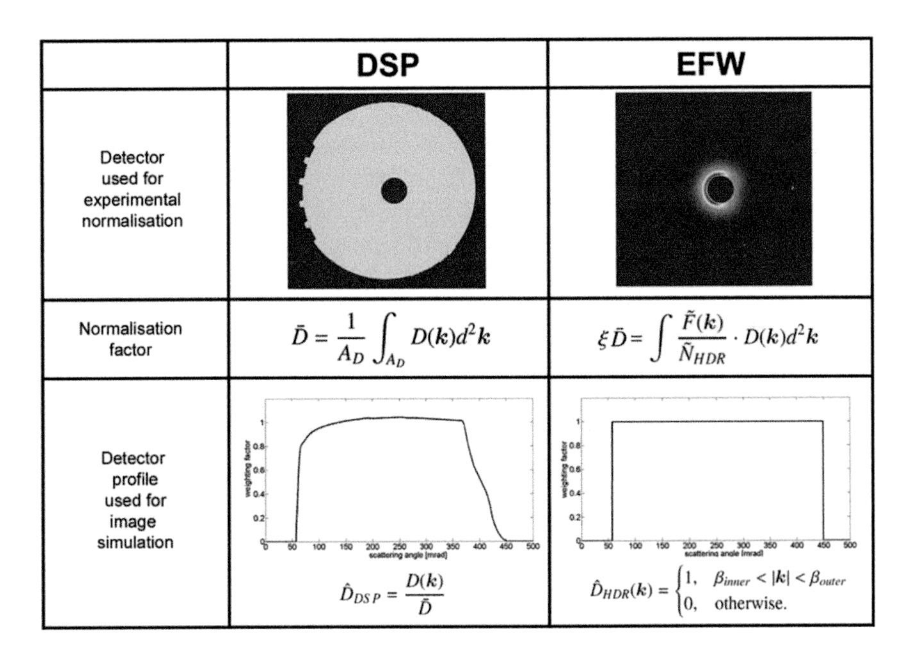

FIGURE 1.11 Summary of the two routes for comparing simulated and experimental quantitative ADF results. Left: The DSP approach where the forward simulations include the detector imperfections. Right: The EFW approach where hardware imperfections are normalized out of the data leaving a result which is not instrument specific and allowing arbitrary simulation codes to be used (Martinez et al. 2015).

1.7 FACTORS AFFECTING THE ACCURACY OF EXPERIMENTAL INTENSITIES AND PEAK POSITIONS

In this section we will discuss factors that can lead separately or cumulatively to errors in the ADF intensity quantification.

1.7.1 Sample Tilt

With the exception of 2D materials and small nano-clusters, to obtain atomic resolution, ADF STEM must be recorded close to low-order crystallographic axes. In the theory of electron channeling, where atoms in aligned crystals act as small lenses for the beam (focusing it for those that follow), we can understand this "wave-guiding" effect as keeping intensity on an atomic column as it propagates through samples far thicker than the probe's depth of field (Van Dyck and De Beeck 1996; Rossouw et al. 2014). The counterpoint then, is that small mis-tilts away from this condition have marked detrimental effects on both image resolution and scattered ADF intensity (Figure 1.12).

Figure 1.12 shows that experimentally, sample tilts of up to 15 mrad can still yield an atomic resolution image (Maccagnano-Zacher et al. 2008) but can lead to a loss of ADF intensity of up to 25% or more (Katz-Boon et al. 2013). In some cases, balancing of the elastic and inelastic contributions to image-intensity can lead to a sample mis-tilt invariant "plateau" in the scattered cross-sections for tilts less than or equal to the convergence angle (MacArthur et al. 2015). This gives the experimentalist a very useful operating range of acceptable mis-tilt and again should be considered in the experiment design.

Sample mis-tilt not only affects image intensity, but it can also affect the apparent position of atomic columns within an image. Often, this effect is neglected; however, where precise peak-position shifts are being measured (oxygen octahedral tilts, cation shifts, etc.) this becomes essential. For samples with sub-lattices of different

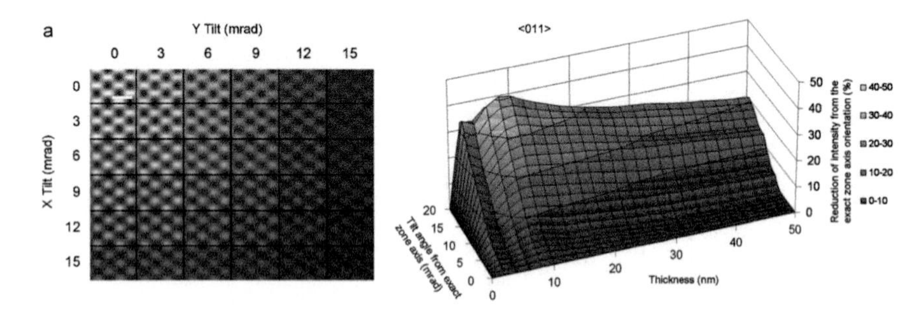

FIGURE 1.12 Left: Simulated ADF images of silicon [110] displayed with fixed contrast limits showing a loss in intensity with increasing tilts (simulation details and image credit, reference (Maccagnano-Zacher et al. 2008)). Right: Quantitative assessment of the loss of scattering for mis-tilted crystals of [110] oriented gold. Here a tilt of only 20 mrad yields a 40% drop in scattering (simulation details and image credit, reference (Katz-Boon et al. 2013)).

composition whose channeling lengths will vary, the same sample tilt may lead to different shifts of apparent peak-positions of columns of differing mass. This error between differing sublattices can be as much as 15 pm for a sample tilted to around 10 mrad (So and Kimoto 2012). The same reference also discusses similar artefacts observed across thickness gradients that are present in a tilted sample. This is relevant for strain mapping of nanostructures, if a reference lattice is defined in a thick region from which lateral strains are measured at a thinner surface region (Yankovich et al. 2014). From the literature we might consider a "rule of thumb" that for column position measurements better than 20 pm, sample tilt must be less than the convergence angle, and for accuracies better than 5 pm, sample tilt should be less than one-half the convergence angle (So and Kimoto 2012).

1.7.2 Atomic-column "Cross Talk"

For thicker specimens, beam-broadening within the sample can lead to scattering reaching the detector originating from several Angstroms to the side of the probe position (Figure 1.13) (Klenov and Stemmer 2006; E et al. 2010). It should be noted that while STEM images are recorded as a function of incident probe position, signal(s) reaching their respective detectors are integrated over the volume of any electron interaction. Figure 1.13 also highlights the improved robustness to the integrated column (or cell) versus the peak intensity.

While larger objective apertures can improve the diffraction limited probe size, de-channeling begins sooner, meaning that optimal experiment design for quantitative ADF requires a trade-off of these considerations (Kotaka 2010; Rossouw et al. 2014; Lefebvre et al. 2015).

The same is of course true for other signal types, and this is directly observable with EDX mapping where de-channeling leads to an artificial widening of otherwise

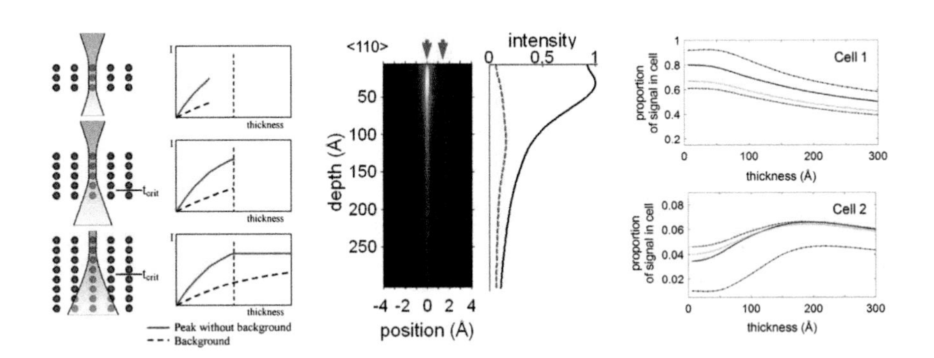

FIGURE 1.13 Left: Schematic of beam broadening and its effect on column and background intensity (figure credit: (Klenov and Stemmer 2006)). Middle: The x-z intensity plot through [110] oriented GaAs sample with the probe positioned over a Ga column (black intensity plot) showing the intensity cross-talk to the nearest As column (dashed red line). Right: Equivalent signal integrated laterally over an atomic-column (cell) for the nearest and neighboring columns (figures modified from (Nguyen et al. 2014)).

sharp interfaces (MacArthur et al. 2017; Spurgeon, Du and Chambers 2017). Again, this is improved when thinner specimens are used but at the expense of total x-ray counts and SNR.

1.7.3 Strain Contrast

Many materials are interesting because of their imperfections such as interfaces, grain boundaries (Fitting et al. 2006), and dislocations (Grillo and Rossi 2011), many of which are accompanied by appreciable strain. These strains distort crystal planes and affect the electron-channeling behavior and hence image contrast; this often increases image intensity in LAADF images at the expense intensity in the HAADF (Figure 1.14).

The precise role of strain in ADF contrast is not yet fully understood, but it is important to be aware of because it is often a manifestation of some other underlying effect such as varying mis-tilt or composition at interfaces (Yu, Muller and Silcox 2004; Wu and Baribeau 2009), each of which would also need to be reflected in any reference simulations.

1.7.4 Amorphous Layers and Carbon Background

For free-standing samples, amorphous layers can often persist with even the most careful sample preparation, whether they are from focused ion beam damage or from carbon contamination from within the microscope itself. Equally for samples supported on thin carbon (or other) membranes, the effect of this "second sample" cannot be ignored (De Backer et al. 2015). In either case, and depending on defocus, they may appear invisible (Figure 1.15) but still contribute to the quantitative image intensity (Mkhoyan et al. 2008; Lefebvre et al. 2015).

In Figure 1.15, though it appears that little resolution is lost by under-focusing through the amorphous layers (last row of panels), this amorphous material remains and will increase beam broadening.

If nanostructures are supported on a thin membrane, it may be possible to improve the quantitative analysis by subtracting some image intensity contribution extrapolated from the region outside the particle (De Backer et al. 2015); however, in other

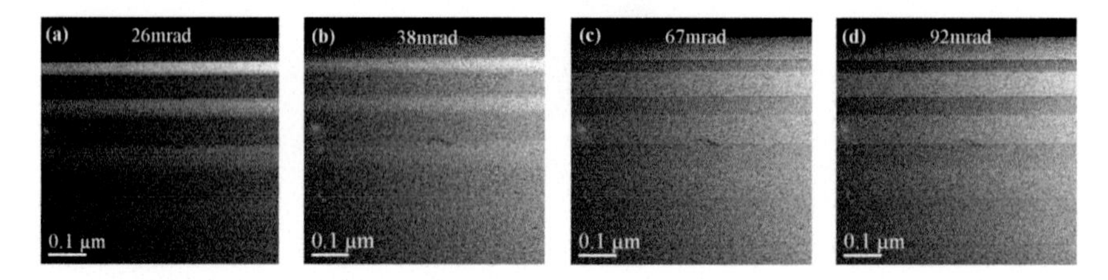

FIGURE 1.14 For larger inner-angles typical z-contrast is observed in HAADF images; however, strain effects can have a very significant role on the contrast of lower angle (LAADF or MAADF) data leading to diminished or even inverted contrast (figure credit: reference (Wu and Baribeau 2009)).

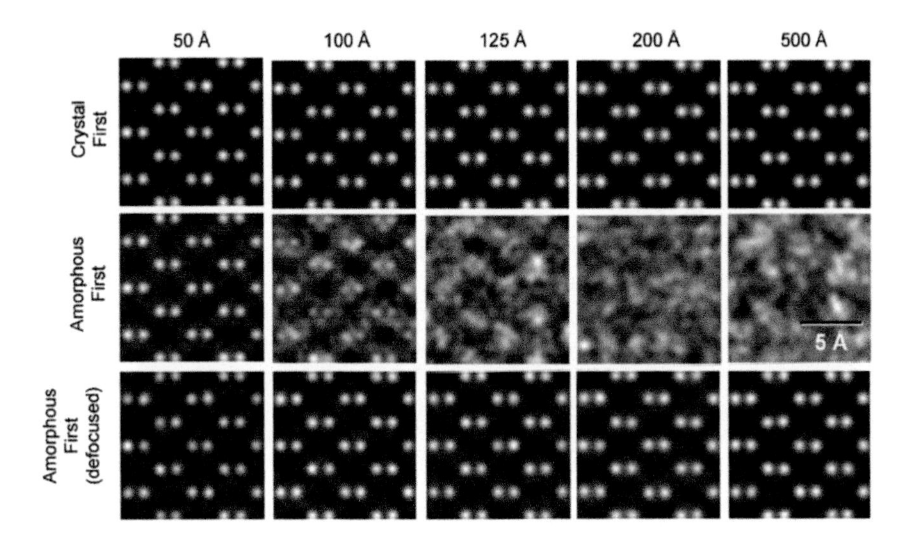

FIGURE 1.15 Simulated data showing the effect of 50% thickness amorphous surface layers on the contrast recorded in ADF STEM for various sample thicknesses. Note that amorphous layers on the rear face of the crystal are invisible at all thicknesses. For the amorphous-first case small underfocus again renders this layer nearly invisible. Full simulation conditions, and figure credit: (Mkhoyan et al. 2008).

situations amorphous carbon may lead to an unavoidable overestimate of the sample thickness (Lefebvre et al. 2015) or a loss of statistical counting precision.

1.7.5 Electron Dose and Pixel Size

The electron beam unfortunately carries the ability to cause sample damage via radiolysis, knock-on, or beam heating (Egerton, Li and Malac 2004). Too much beam intensity and our sample will become transformed hindering the precise measure we are trying to obtain; however, too little beam intensity and the low signal-to-noise caused by excessive shot noise may make quantitative measurements less precise (Van Aert et al. 2019). Statistical studies are useful for predicting the expected error as a function of dose for different applications such as nanometrology (Van Aert et al. 2013; De Backer et al. 2015; De Backer et al. 2015), or light element imaging (Hovden and Muller 2012; Gonnissen et al. 2014). Recall from Figure 1.9 that for a given fixed dose, the expected probability of error is always bigger for peak-intensity measurements compared with integrated cross-sections. It is often perceived that finer pixel sampling gives "better" pictures; however, these may contain no additional quantitative information content. As long as the sampling is sufficiently high to meet the instrument resolution (say equal or better than twice Nyquist) then additional pixels do not improve the reliability of measurements (E et al. 2013; De Backer et al. 2015). Up to 4x Nyquist may be useful when onward reregistration or processing is planned but again, beyond that no additional benefit is conveyed.

1.7.6 Cold Field-emission Current Fluctuation and Emission Decay

In general, emission fluctuations from cold field-emission type electron sources are monitored within the gun and are compensated in real time. Where this fails, small fluctuations of a few percent in image intensity may result, and users should be aware of this when using such systems. This is relatively easy to spot in bright-field or annular bright-field (ABF) data if an area of vacuum is included in the field of view (as a totally featureless region would be expected), but it is far harder to observe in ADF data. For this reason, and where it is available, it is useful to record the bright-field or ABF image in addition to the ADF data. One solution to this, is to monitor the real-time variation of beam current, and to incorporate this in the ADF normalization (House, Schamp and Yang 2018). Perhaps more importantly with these cold field-emission gun (FEG) systems is to be aware of any emission current decay between the detector mapping and the experimental acquisition (Jones et al. 2014).

1.7.7 Summary of Sources of Error

Table 1.1 Overleaf summarizes the dominant sources of relative error in quantitative ADF ranked by severity if not properly mitigated. Not all will apply in all instances and magnitudes will vary with individual systems and experiments.

1.8 ENVIRONMENTAL NOISE AND SCANNING-DISTORTION IN THE STEM

As the resolution and magnification of modern STEMs has improved, so too has their sensitivity to the laboratory environment. This sensitivity can be to acoustic, thermal, electromagnetic, or seismic disturbances (Muller and Grazul 2001; Muller et al. 2006). Environmental instability can affect the quality of scanned image data dramatically (Muller et al. 2006; Jones and Nellist 2013).

One approach is to record multiple frames and realign these by cross-correlation (Kimoto et al. 2010; Couillard et al. 2013; Spurgeon et al. 2015); in each case after averaging these data both improve the SNR and yields a column-position precision of several picometers or better (Kimoto et al. 2010; Zuo et al. 2014; Sang et al. 2015; Aso et al. 2016). However, rigid registration and averaging may not mitigate residual low-frequency time-varying environmental distortion, and artefacts often remain visible in strain maps (Zuo et al. 2014; Couillard et al. 2013).

To understand the way in which scan artefacts affect serial-recorded data, we must understand the frequency domain over which any environmental disturbance operates [see (Jones et al. 2015) for a more detailed discussion of frequency ranges]. For data recorded sufficiently above the Nyquist sampling limit, high-frequency scan-noise in the kilohertz range may corrupt peak-positions and the peak-intensities of individual pixels; however, it does not affect the *integrated intensity* over an atomic column. The same cannot be said of low-frequency scan-distortions (approximately 0.5- to 3-Hz range), and the effect of these on quantitative ADF image intensities remains an important area of research (Jones et al. 2015). To understand the difference in behavior between *scan-noise* and *scan-distortion*, we must consider the nature of the unwanted STEM probe offsets in each case. For high-frequency scan-noise, the probe is quasi-randomly offset leading to individual scan lines through features being affected (sometimes called variously "skipping", "tearing", or "flagging"). As the scan lines are affected individually, and because the offsets are random in nature, there is no systematic way to say that peak-positions or intensities are biased higher or lower; but rather they are more scattered and less precise.

On the other hand, scan-distortion is observable at far slower frequencies and an ensemble type effect is observable where several neighboring scan lines are stretched, squashed, or sheared collectively (Sun and Pang 2006). This has two effects. First, peak-positions are translated by the collective effect of this lingering probe offset, which acts across the multiple scan lines. Second, the integrated intensity of features is affected. This can be better understood if we consider the vector field of the probe offsets, Figure 1.16 (Jones et al. 2015).

Once the probe offsets are considered as a vector field, we can consider the fast- and slow-scan directions separately; if we consider the error in the slow-scan direction varying with slow-scan direction (its gradient), we see this represents the stretching or squashing of the image in local regions. While this may not affect the

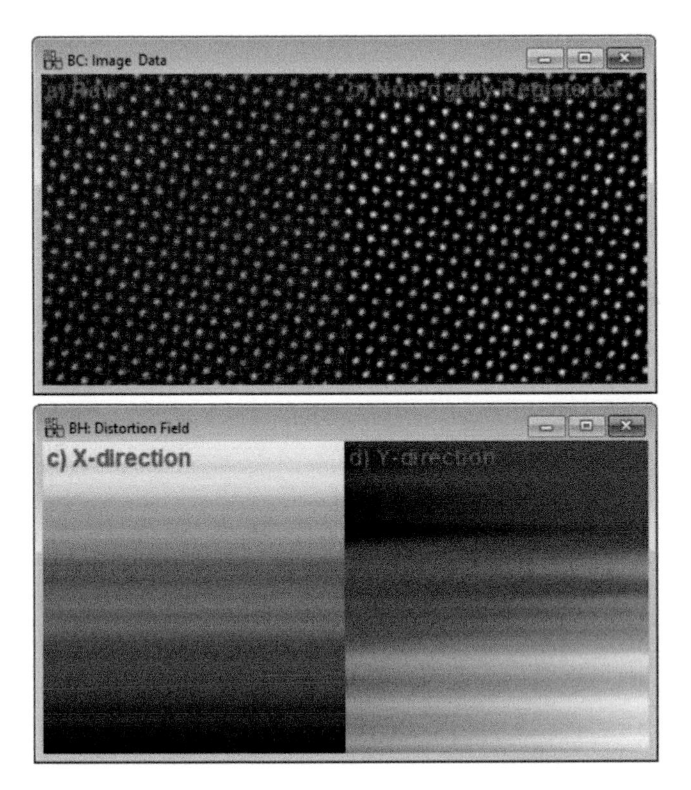

FIGURE 1.16 Examples of x and y translation data for a non-linear non-rigidly registered frame. The image (a) is non-rigidly registered and produces the corrected frame (b) after applying probe offsets to correct the diagnosed errors in the x-direction (c) and the y-direction (d).

TABLE 1.1
Dominant sources of error in quantitative ADF image intensities ordered by severity if left unmitigated.

Error Source	Severity	Cause	Solution	Comments
Detector asymmetry	Global error, up to 15 % if ignored with worst-case detector (Martinez et al. 2015)	Detector geometry/design (MacArthur et al. 2014)	Improve design/symmetry (Kaneko et al. 2014)	Can be as low as 2 % if corrected during normalization or simulation (Findlay and LeBeau 2013; Martinez et al. 2015)
Sample tilt	Local error, up to 10 % for peak-intensities, ≈ 2 % for cross-sections, at 5 mrad mis-tilt (E et al. 2013; Katz-Boon et al. 2013)	Loss of electron channeling condition	Optimize convergence angle to be more robust to tilt (MacArthur et al. 2015), keep tilt $\leq 0.5\ \alpha$	Varies by column, depends on sample thickness
Scan-flyback hysteresis	Spatially varying, can be as bad as 15–20 % at leading edge of image	Finite inductance (response rate) of STEM scan coils (Buban et al. 2010)	Increase flyback-waiting time (typically > 600 µs), or correct in postprocessing	Other nonstandard scan paths exist (Sang et al. 2016)
Detector angle mis-measurement	Global error, ≈ 1 % error per 1 % error in inner-angle, ≈ 0.2 % error per 1 % error in outer angle	Errors lead to incorrect comparison simulations	Careful detector angle measurement, including outer-angle/flux cut-off evaluation (LeBeau and Stemmer 2008; Martinez et al. 2015; Yamashita et al. 2015)	Camera-length dependent; statistical analysis approaches not affected (Van Aert et al. 2011)
ADF response temporal effects	Global error, 2-10 % if different dwell-times used for ADF scan (Krause et al. 2016; Sang and LeBeau 2016) Local error, image streaking if dwell-time is ≤ 2 µs (Sang and LeBeau 2016)	PMT and read-out electronics response time (Buban et al. 2010)	Use dwell-times > 2 µs Use same dwell-time for ADF scan	Be aware of acquisition software's dwell-time scaling behavior

Detector or flux "cut-off" mis-alignment	Global error, up to 3 % with shifts of around 10 mrad (Findlay and LeBeau 2013)	Incorrect projection lens settings or de-scan calibration; incorrect image-shift setting	Improve column alignment verify anode wobble center in Ronchigram	May also cause appearance of astigmatism/peak-shifts (Findlay and LeBeau 2013)
Magnification (pixel width)	Global error, generally < 2 %; cross-section error scales with mag error squared (E et al. 2013)	Imperfect scan coil calibration	Verify calibrations (in x and y) from bulk reference material	Statistical intensity analysis approaches not affected (Van Aert et al. 2011)
Scan noise/distortion	Local error, ≈ 0-5 %, frequency and dwell-time dependent (Jones and Nellist 2013; Jones et al. 2015)	Environmental instability (Muller et al. 2006)	Room design (Muller et al. 2006), fast-scan, and post-processing techniques (Kimoto et al. 2010; Couillard, Radtke and Botton 2013; Ishikawa et al. 2014; Jones et al. 2015; Ophus, Ciston, and Nelson 2016)	May also affect peak-positions and strain mapping (Couillard et al. 2013; Zuo et al. 2014)
Shot noise	Global error, generally < 1 % for cross-sections, worse for peak intensities	Fundamental limit of Poisson distributed intensities (De Backer et al. 2015)	Increase beam current, dwell-time, or number of frames (Katz-Boon et al. 2013)	Error scales with 1/√dose (Katz-Boon et al. 2013)
Amplifier brightness/contrast problems	Local error depending on sample thickness, < 1 % if amplifier adjusted properly (Grillo 2011; Ishikawa et al. 2014)	Poisson distribution of counts can lead to "preclipping" in thick sample regions (Grillo 2011)	Observe waveform during acquisition, increase brightness, or reduce amplifier gain	Up to 10 % if brightness is too low or gain is too high (Grillo 2011)
Emission current fluctuation	Local error, generally < 1 %, affects bands parallel to fast-scan	Power supply and tip stability	Postprocessing (House et al. 2018), or fast-scan averaging techniques (Kimoto et al. 2010; Couillard et al. 2013; Jones et al. 2015)	Cold-FEGs only; error may become 5–10 % if compensation fails

peak intensity, it is clear that a compression will reduce the intensity of an integrated ADF cross-section and vice versa; an expansion will increase the integrated intensity of a feature. To correct for this slowly varying (and often nonlinear) distortion there are two families of approaches depending on whether or not prior knowledge is introduced.

If prior knowledge is introduced, say by imposing known crystal-plane angles, we can correct for linear (thermal) stage drift but not for non-linear effects (Jones and Nellist 2013). Alternatively, fixing atomic-column positions to those from a previously recorded TEM image (Rečnik, Möbus and Sturm 2005), or to a regular grid (Wang et al. 2018), may improve lattice appearance but risks washing away genuine defect or local-strain information.

Perhaps a more well-established set of approaches are those that avoid as much as possible introducing sample-specific prior knowledge. Of course, these also are built on other implicit assumptions. Approaches in which the scan rotation is precessed (followed by affine correction of lattice angles) rely on having a fixed or very slowly decaying monotonic drift vector (Sang and LeBeau 2014; Fujinaka et al. 2020); this is reasonable for slowly subsiding thermal drift and can fully correct linear drift without crystal prior knowledge. An alternative (and equally valid) assumption is to acknowledge that the slow-scan speed is typically around two to three orders of magnitude slower than the fast-scan time. This can then be utilized either in special cases where some perfect reference crystal is visible in part of the field (Wen et al. 2010), or more generally for arbitrary samples by recording an orthogonal pair of images and attributing reliable lattice distances to each one along each image's fast-scan direction. Any distortions in the slow-scan direction are non-linearly warped to the fast scan of the accompanying frame (Ophus, Ciston and Nelson 2016).

In recent years, using non-rigid techniques, peak-position precision has improved to approximately 1 pm levels (Yankovich et al. 2014; Gonnissen et al. 2016). It may be possible to further improve this in the coming years or there may be some as yet undiscovered lower bound to the attainable precision.

1.9 FURTHER MULTI-FRAME APPLICATIONS OF STEM IMAGING, SPECTROSCOPY, AND 4D-STEM

In this section, some recent applications of multi-frame data acquisition and processing are discussed.

1.9.1 Increasing Image SNR

The most direct use for multi-frame averaging is to increase the SNR from individual frames. Figure 1.17 shows an example of such an averaged image from a relatively beam-sensitive gold nanoparticle.

If ADF data were purely Poisson limited, the expected SNR improvement would scale with the square root of the number of frames; for example, averaging just 25 frames would be expected to increase SNR 5x. Some important observations stem from this relationship. First, though very useful, there will always be a diminishing

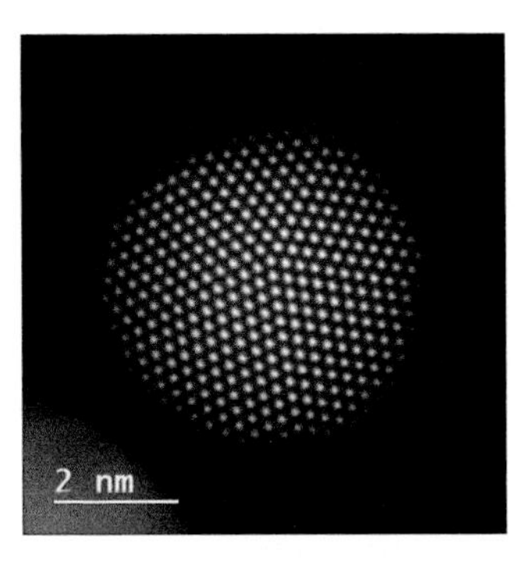

FIGURE 1.17 Non-rigid aligned image from 70 fast-scanned ADF frames of an Au-decahedron. The large fraction of surface atoms and metastable twinning makes this a relatively beam-sensitive specimen. Data was recorded at the David Cockayne Centre for Electron Microscopy, University of Oxford, and realigned using the SmartAlign algorithm (Jones et al. 2015).

return from imaging ever-increasing numbers of frames. The square-root-n type behavior of the SNR improvement means that doubling the number of frames leads to only around 40% more SNR. If increasing the number of frames introduces parts of the acquisition for which the sample was damaged, tilting, or for which the microscope tuning (e.g. focus or astigmatism) drifted very far, then these extra frames may eventually make the data less precise overall. The skill in designing multi-frame experiments and the subsequent data analysis is often to spot these deteriorations and to only average the frames up until this point.

The second observation is that not all ADF data is Poisson limited. For relatively high doses, this is a very reasonable assumption; however, when the dose becomes incredibly low such that individual electrons become visible on top of a non-trivial dark-noise, this same assumption does not hold and other software or hardware data pre-treatments may be necessary (Jones and Downing 2018; Mittelberger, Kramberger and Meyer 2018).

1.9.2 Increasing Spectroscopic SNR

In the same way as image alignment and summation, spectroscopic signal accumulation is a powerful way to increase data SNR. There are two approaches to achieve this depending on the way the spectrum-volume data series are handled.

For every energy channel in the spectrum volume, it is possible to extract its own frame series; thus a spectral volume with (say) 2048 energy channels can be reshaped to become 2048 movies with the same number of frames as the original time-series (Yankovich et al. 2016). On completion the distortion-corrected data are summed then reshaped again to a single spectral volume with the original dimensionality. This "channel preference" approach is perhaps useful if only a very small part of the spectrum is of interest (for example, perhaps just one EDX peak), but it requires every spectrum to be opened from the disk to create each series before it can be processed.

An alternative approach is to undistort each spectral volume one at a time after loading into memory, and to add these to a cumulative running-total volume (Jones et al. 2018a; Wang et al. 2018). This approach has the advantage that the memory used by the computer is only ever the raw single-scan, corrected single-scan, and the cumulative-total data volumes, regardless of the number of frames being processed overall.

Mathematically both approaches should be able to achieve similar results and only differ in computing approach. Memory efficient methods are essential when this concept is extended to the compensation of multi-frame four-dimensional (4D) data.

Template matching can be used to further improve the SNR of periodically repeating structures, though necessarily this is not appropriate for analyzing a single unique point defect. Such template matching may be large regions of material (such as in Figure 1.18), patches of large unit-cells (Jones et al. 2018a), small motifs from ordered materials (Wenner et al. 2017), or even single atomic columns (Jeong and Mkhoyan 2016).

There are also some experimental indications that sharing a fixed total electron dose across several scan frames impedes some damage mechanisms (Jones et al. 2018a). However, although dose-fractionation and summing presents many

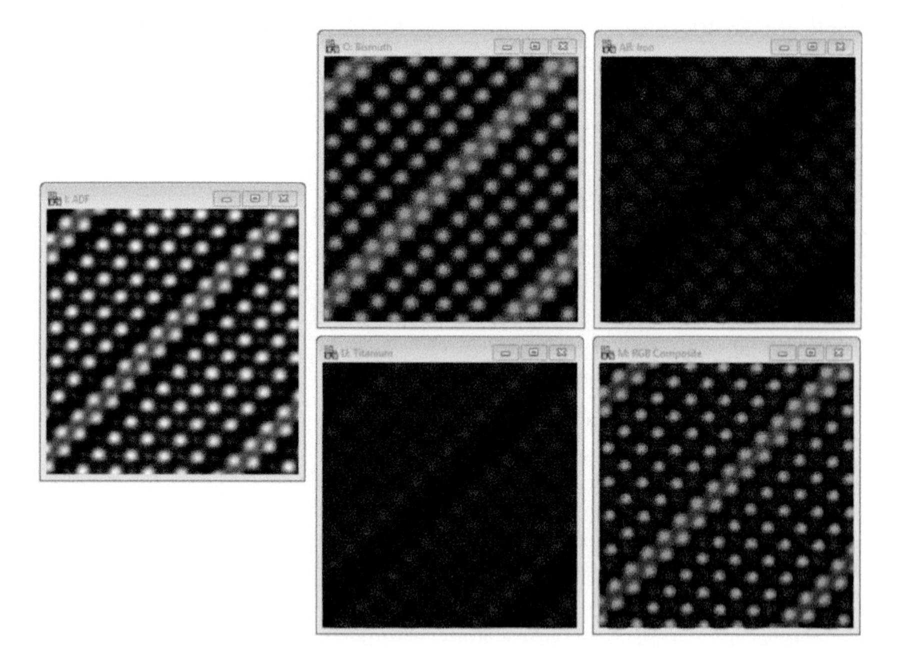

FIGURE 1.18 Atomic resolution example of combined multi-frame ADF + EELS + EDX simultaneous acquisition with subsequent template-matching (region of interest ~4 nm). The sample is an Aurivillius phase $Bi_6Ti_xFe_yMn_zO_{18}$ and is a rare example of a room temperature multiferroic. Raw data comprised 21 experimental scan frames recorded with a dwell-time of 20 ms and a beam current of 25 pA. Lighter elements iron (red) and titanium (blue) were mapped using EELS, while the heavier bismuth (green) was mapped using EDX (sample courtesy Dr. Lynette Keeney, further details (Keeney et al. 2017)).

opportunities to improve scan fidelity and reduce beam damage, it is not possible to divide this dose endlessly. For EDX detectors (single photon sensitive digital counters) with a perfect Poisson noise limitation, frame summing yields excellent results even where hundreds of frames are summed (Jones et al. 2018a). However, for EELS spectrometers with a non-trivial dark noise the optimum performance is often found with around 5–10 frames. In the future these various multiframe spectroscopic tools may be possibly usefully combined with other novel signal collection strategies, especially for EELS (Kimoto et al. 2005; Bosman and Keast 2008; Haruta et al. 2019), or for electron magnetic circular dichroism (EMCD) where the reliability of conclusions from single-frame scans is just about within reach (Negi et al. 2018; Negi et al. 2019).

1.9.3 Increasing Pixel Density (Digital Super-resolution)

Digital super-resolution is a relatively recent area of research in STEM imaging and is another approach in the scanning-design toolkit (along with denoising (Mevenkamp et al. 2015; Spiegelberg et al. 2018) or sparse-scanning with inpainting (Anderson et al. 2013; Stevens et al. 2014)).

Although denoising and inpainting appear attractive venues at first, most denoising introduces some element of filtering, which may result in SNR gain at the expense of resolution (sharpness), or in the case of block-matching, a reliance on prior knowledge. Sparse scan approaches also require complex fast deflector or blanking systems (Béché et al. 2016; Kovarik et al. 2016) Very recently, some groups have made quantitative assessments of such techniques, and especially in the case of sparse scanning and inpainting, have shown these to be of limited use for image improvement (Sanders and Dwyer 2018; Van den Broek et al. 2019). Sparse sampling, however, remains useful where the goal is to reduce dose exposure or data rates or to increase video frame rates (Stevens et al. 2018).

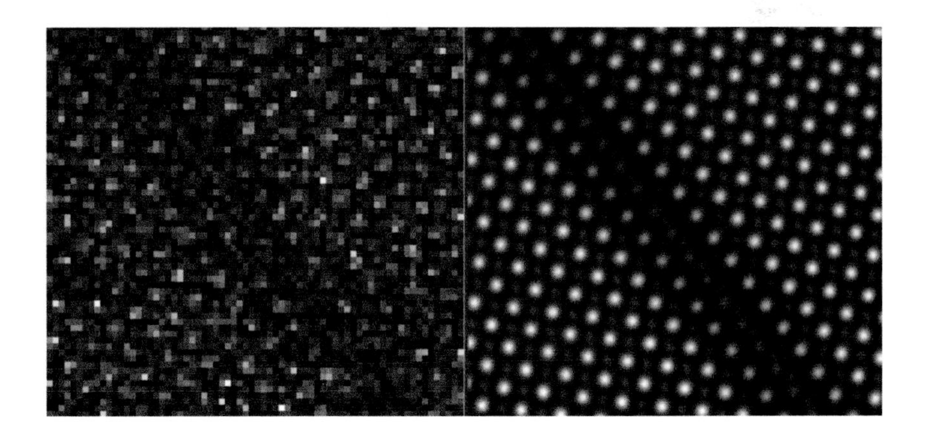

FIGURE 1.19 Example of 4x upsampling during super-resolution reconstruction. Data originally consisted of 20 frames recorded at 2x Nyquist (image width ~6 nm). While the original data at 2x Nyquist mut have contained the spatial information encoded within it, it is not directly interpretable to the user until the frame series is reconstructed to the final upsampled average.

At its most simple, digital super-resolution (as distinct from optical super-resolution) is the practice of fusing multiple lower quality images to produce one higher quality image (Jones et al. 2015; Bárcena-González et al. 2017, Bárcena-González et al. 2019). Uses of this include extending image resolution and improving the quantification of strain mapping (Bárcena-González et al. 2016), and because of this the overlap with multi-frame non-rigid registration is obvious.

1.10 FUTURE POSSIBILITIES IN QUANTITATIVE ADF AND MULTIFRAME IMAGING

In this final section, some recent approaches and possible future direction for quantitative ADF imaging are discussed.

1.10.1 New Geometries of Dark-field Detection

The basic geometry of the ADF detector has changed little since its invention; however, several possibilities exist for the future of quantitative ADF with some recently developed detectors. Figure 1.7 (Detector E) shows perhaps the most simple of these—the "quadrant-detector" ADF—where the annulus has been divided into four quarters. At shorter camera-lengths these can be added together to give a classical ADF image or, at longer camera-lengths, the signals from opposite sides can be subtracted to give a differential phase-contrast signal (Shibata et al. 2012). This has obvious advantages in hardware flexibility, but as Figure 1.7 shows, the correct setup of each individual amplifier now becomes crucial. Taking this concept further, multiple annular rings (themselves each divided into quadrants) can be used (Figure 1.20) (Shibata et al. 2010).

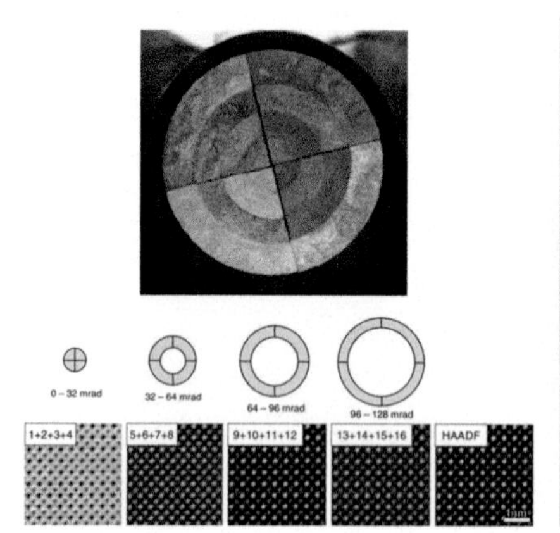

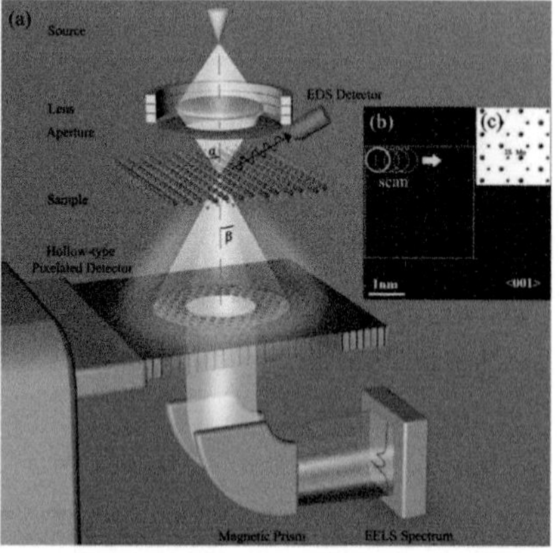

FIGURE 1.20 Left: Example of a fiberoptically coupled 16-segment photomultiplier-style detector and some available geometries and their associated images. A simultaneous conventional HAADF image is also shown [figure modified from reference (Shibata et al. 2010)]. Right: A recent proposal for a pixelated detector with a central through-hole to enable EELS spectroscopy [figure credit: (Song et al. 2018)].

Such a detector allows for a host of sum, or difference, images to be synthesized after the scattering data have been recorded, potentially moving experiment design from a pre-imaging to a post-processing regime. These detectors seem ideally suited to imaging of light elements (Ooe et al. 2019), and differential phase-contrast (Shibata 2019) and electric-field imaging (Hachtel, Idrobo and Chi 2018) has now been demonstrated even up to atomic resolution, and discussions about the scope of this new technique are ongoing (Clark et al. 2018).

Taking this concept to its extreme is the use of pixelated detectors, which can be used to record all the scattering in the detector plane (Figure 1.21). This yields flux-patterns similar to Figure 1.10, including Kikuchi patterns for thicker samples (Kimoto and Ishizuka 2011), but it has the advantage of a homogenous detection efficiency through applying conventional dark- and gain-reference approaches.

With CCD technology dynamic range is an issue, making it difficult to record both the bright-field and dark-field regimes. In Figure 1.21 a beam-stopper was used to mask the bright-field disk; however, this contains a great deal of useful sample information (Pennycook et al. 2015). More recently, dedicated detectors with large dynamic ranges have been developed to overcome this (Tate et al. 2016). An expansive discussion of all the new models of pixelated STEM sensors is outside the scope of this chapter, but a recent review by Ophus (2019) provides far more insight. Although these pixelated detectors offer more flexibility by recording the whole Ronchigram, they are considerably more expensive, far slower than regular ADF scanning rates (and hence more sensitive to environmental distortion), and can lead to a "data deluge" where the microscopist also must become a data-scientist. Alternatives, such as the position-sensitive detector, may represent a desirable "sweet spot" of additional phase-contrast, magnetic field, or electric-field information, without the complexity of full 4D-STEM (Schwarzhuber et al. 2018). There also exist possibilities for detectors in other planes (Lazar et al. 2011) or for the imaginative use of apertures to approximate this 4D-STEM (Müller-Caspary et al. 2016).

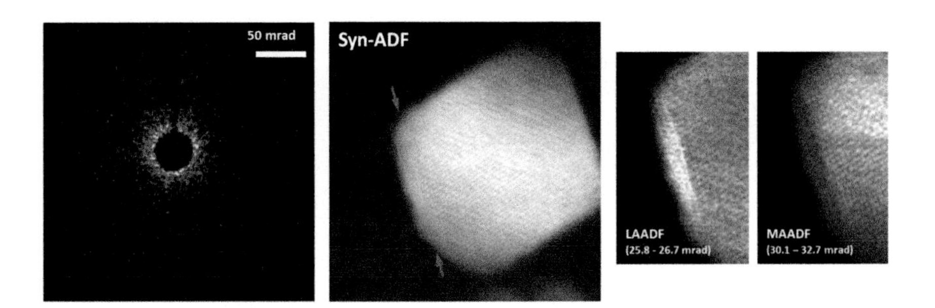

FIGURE 1.21 Left: Example of pixelated detector read-out. A beam-stopper was used to mask the bright-field disk. Bright points indicate individual electron-scattering events. Right: Example images synthesized from the 4D dataset, the integrated ADF signal (arrows indicate the position of a twin boundary), and synthetic LAADF and MAADF images where the edge region appears bright and dark, respectively (figure credit see ref:(Jones 2016)).

1.10.2 Single Electron Counting Binary ADF Imaging

Pursuing ever lower electron doses is an active research area for beam-sensitive specimens. Recently, several groups have observed that at ultra-low doses individual electron scattering events to the ADF detectors become visible (Figure 1.22) (Krause et al. 2016; Sang and LeBeau 2016; Jones and Downing 2018; Mittelberger, Kramberger and Meyer 2018).

In all these cases, a sharp pulse onset is observed, and for some a far slower exponential-like decay follows (Foord et al. 1969). This decay appears to be the true cause of reported image-streaking in ultra-short dwell-time frames previously misidentified as scan-noise (Buban et al. 2010). This quantized nature of the electron pulses invites the possibility of a digital electron-counting approach to reduce detector efficiency inhomogeneity (Findlay and LeBeau 2013; MacArthur, Jones and Nellist 2014), and to wholly eliminate dark-noise. This approach may also be useful as part of a future avenue to circumvent some types of radiation damage (Meyer, Kotakoski and Mangler 2014).

1.10.3 Advances in Scan patterning and Custom Scan-Design

By far, the vast majority of STEM data are captured using a simple Cartesian raster. Strictly, the actual implementation in most systems is not quite a perfect raster scan

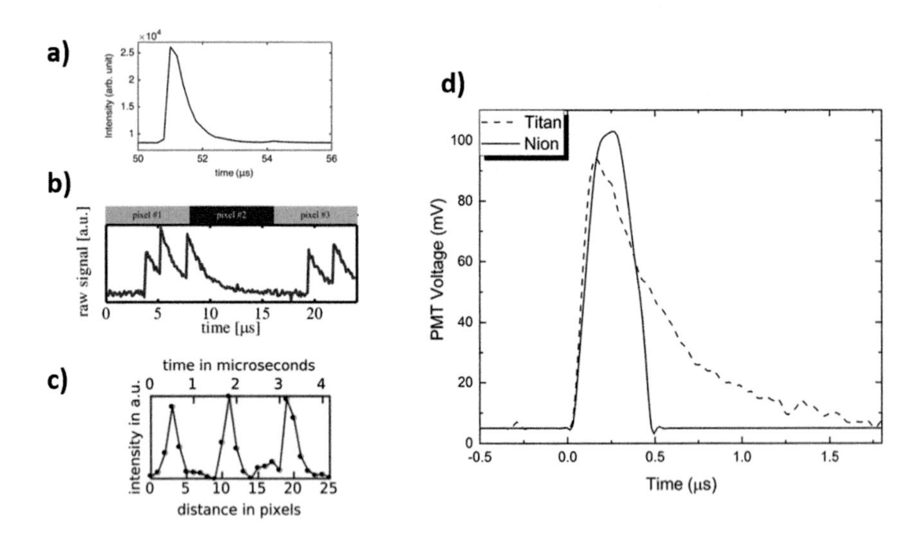

FIGURE 1.22 A collection of literature results indicating the potential for electron counting with current-generation ADF hardware. (a) A coarsely sampled single electron-pulse captured from a Titan STEM (Sang and LeBeau 2016). (b) A more finely sampled, but multiple pile-up recording of sequential pulses again on a Titan STEM (Krause et al. 2016). (c) A coarsely sampled set of three electron-pulses recorded using a Nion STEM (Mittelberger et al. 2018). (d) A finely sampled explicit comparison of these two manufacturers (Jones and Downing 2018). Note the differing pulse shape and width resulting from the differing tuning of the onboard PMT amplifiers.

pattern, as often significant flyback is added to the start of every fast-scan line. When flyback time is reduced significantly, a quasi-exponential distortion is clearly visible at the start of each line (usually the left edge of an image) (Anderson et al. 2013; Sang et al. 2016).

Another idea which arises in the literature periodically is the use of spiral scans (Wang et al. 2010; Sang et al. 2016). These do present some advantages for prior knowledge free compensation of linear drift, but they are generally hard to implement for the average user, introduce inhomogeneous electron-dose density, or result in circular images less favored for publication or communication.

Other innovative scanning approaches include line-hopping (Kovarik et al. 2016), Lissajous scans (Li et al. 2018), or approximately random sampling (Zobelli et al. 2019). Again, these are non-trivial to implement, but they are somewhat useful hybrids of sparse sampling (from which it is impossible to ascertain unwanted probe-offsets) and raster sampling. These approaches deserve further study as significant improvements in frame-rate are achievable if other issues can be resolved.

1.11 CONCLUSION

Quantitative ADF has seen a resurgence in interest over the last fifteen years inspiring a rapid pace of development across both hardware and software. These advances have been further leveraged by image simulation tools that enable experiment planning in advance of recording data, and deep data-mining techniques to analyze the results obtained. Already this has allowed several nanotechnology problems to be investigated. In the coming years the development of quantitative cross-sections for analytical signals such as EELS (Rez 2001) and EDX (Rez 2003; Nguyen et al. 2014; Zhu and Dwyer 2014) is expected, and the incorporation of these offer the potential to unlock the full material characterization of arbitrary samples. The efforts needed to calibrate the instrumentation to deliver truly quantitative signals is not trivial. However, if we are to go beyond merely taking "pretty pictures" and to harness the full power of the aberration-corrected STEM, then these efforts are essential. The literature referenced in this chapter should hopefully point the reader to the efforts already made on their behalf to accelerate their journey, but more than likely the most exciting results are those yet to come.

ACKNOWLEDGMENTS

The author would like to thank Hidetaka Sawada for assistance developing the swung-mapping method and Figure 1.6; Gerardo Martinez, Armand Béché, Quentin Ramasse, and Judy Kim for assistance with Figure 1.10; Aakash M Varambhia for Figure 1.17; and pnDetector for the camera used in Figure 1.21. Some aspects of this research were supported by EU-FP7 Grant 312483 (ESTEEM2), EPSRC Grant EP/K040375/1 (South of England Analytical Electron Microscope), and SFI Grant 12/RC/2278 (AMBER).

REFERENCES

Altantzis, T., Lobato, I., De Backer, A., Béché, A., Zhang, Y., Basak, S., Porcu, M., Xu, Q., Sánchez-Iglesias, A., Liz-Marzán, L.M. and Van Tendeloo, G., 2019. Three-dimensional quantification of the facet evolution of Pt nanoparticles in a variable gaseous environment. *Nano Letters*, *19*(1): 477–481.

Anderson, H.S., Ilic-Helms, J., Rohrer, B., Wheeler, J. and Larson, K., 2013, February. Sparse Imaging for Fast Electron Microscopy. In *Computational Imaging XI* (Vol. 8657, p. 86570C). International Society for Optics and Photonics.

Aso, K., Shigematsu, K., Yamamoto, T. and Matsumura, S., 2016. Detection of picometer-order atomic displacements in drift-compensated HAADF-STEM images of gold nanorods. *Journal of Electron Microscopy*, *65*(5): 391–399.

Bals, S., Van Aert, S., Romero, C.P., Lauwaet, K., Van Bael, M.J., Schoeters, B., Partoens, B., Yücelen, E., Lievens, P. and Van Tendeloo, G., 2012. Atomic scale dynamics of ultrasmall germanium clusters. *Nature communications*, *3*(1): 1–6.

Bárcena-González, G., Guerrero-Lebrero, M.P., Guerrero, E., Yañez, A., Fernández-Reyes, D., González, D. and Galindo, P.L., 2017. Evaluation of high-quality image reconstruction techniques applied to high-resolution Z-contrast imaging. *Ultramicroscopy*, *182*: 283–291.

Bárcena-González, G., Guerrero-Lebrero, M.P., Guerrero, E., Yañez, A., Nuñez-Moraleda, B., Kepaptsoglou, D., Lazarov, V.K. and Galindo, P.L., 2019. HAADF-STEM image Resolution enhancement using High-quality image reconstruction techniques: case of the Fe 3 O 4 (111) surface. *Microscopy and Microanalysis*, *25*(6): 1297–1303.

Bárcena-González, G., Lebrero, M.P., Guerrero, E., Reyes, D., González, D., Mayoral, A., Utrilla, A.D., Ulloa, J.M. and Galindo, P.L., 2016. Strain mapping accuracy improvement using super-resolution techniques. *Journal of Microscopy*, 262(1): 50–58.

Béché, A., Goris, B., Freitag, B. and Verbeeck, J., 2016. Development of a fast electromagnetic beam blanker for compressed sensing in scanning transmission electron microscopy. *Applied Physics Letters*, *108*(9): 093103.

Bosman, M. and Keast, V.J., 2008. Optimizing EELS acquisition. *Ultramicroscopy*, *108*(9): 837–846.

Buban, J.P., Ramasse, Q., Gipson, B., Browning, N.D. and Stahlberg., H. 2010. High-Resolution Low-dose scanning transmission electron microscopy. *Journal of Electron Microscopy 59*(2): 103–12. https://doi.org/10.1093/jmicro/dfp052.

Clark, L., Brown, H.G., Paganin, D.M., Morgan, M.J., Matsumoto, T., Shibata, N., Petersen, T.C. and Findlay., S.D. 2018. Probing the limits of the rigid-intensity-shift model in differential-phase-contrast scanning transmission electron microscopy. *Physical Review A, 97*(4): 043843. https://doi.org/10.1103/PhysRevA.97.043843.

Couillard, M., Radtke, G. and Botton, G.A., 2013. Strain fields around dislocation arrays in a Σ9 silicon bicrystal measured by scanning transmission electron microscopy. *Philosophical Magazine, 93*(10–12): 1250–67. https://doi.org/10.1080/14786435.2013.778428.

Craven, A.J., Sawada, H., McFadzean, S. and Maclaren., I. 2017. Getting the most out of a post-column EELS spectrometer on a TEM/STEM by optimising the optical coupling. *Ultramicroscopy*, *180*(September): 66–80. https://doi.org/10.1016/j.ultramic.2017.03.017.

De Backer, A., Gonnissen, J. and Van Aert, S., 2015. Optimal experimental design for nanoparticle atom-counting from high-resolution STEM images. *Ultramicroscopy*, *151*: 46–55.

De Backer, A., Jones, L., Lobato, I., Altantzis, T., Goris, B., Nellist, P.D., Bals, S. and Van Aert, S., 2017. Three-dimensional atomic models from a single projection using Z-contrast imaging: verification by electron tomography and opportunities. *Nanoscale*, *9*(25): 8791–8798.

De Backer, A., Martinez, G.T., MacArthur, K.E., Jones, L., Béché, A., Nellist, P.D. and Van Aert, S., 2015. Dose limited reliability of quantitative annular dark field scanning transmission electron microscopy for nano-particle atom-counting. *Ultramicroscopy, 151*: 56–61.

De Backer, A., Martinez, G.T., Rosenauer, A. and Van Aert, S., 2013. Atom counting in HAADF STEM using a statistical model-based approach: methodology, possibilities, and inherent limitations. *Ultramicroscopy, 134:* 23–33.

De Backer, A., Van den Bos, K.H.W., Van den Broek, W., Sijbers, J. and Van Aert, S., 2016. StatSTEM: an efficient approach for accurate and precise model-based quantification of atomic resolution electron microscopy images. *Ultramicroscopy, 171*: 104–116.

De wael, A., De Backer, A., Jones, L., Nellist, P.D. and Van Aert, S., 2017. Hybrid statistics-simulations based method for atom-counting from ADF STEM images. *Ultramicroscopy, 177*: 69–77.

De wael, A., De Backer, A., Jones, L., Varambhia, A., Nellist, P.D. and Van Aert, S., 2020. Measuring dynamic structural changes of nanoparticles at the atomic scale using scanning transmission electron microscopy. *Physical Review Letters, 124*(10): 106105.

E, H., MacArthur, K.E., Pennycook, T.J., Okunishi, E., D'Alfonso, A.J., Lugg, N.R., Allen, L.J. and Nellist, P.D. (2013). Probe integrated scattering cross sections in the analysis of atomic resolution HAADF STEM images. *Ultramicroscopy, 133*: 109–119. http://www.sciencedirect.com/science/article/pii/S0304399113001770 (Accessed July 16, 2013).

E, H., Nellist, P. D., Lozano-Perez, S. & Ozkaya, D. (2010). Towards quantitative analysis of core-shell catalyst nano-particles by aberration corrected high angle annular dark field STEM and EDX. *Journal of Physics: Conference Series* 241, 012067. http://stacks.iop.org/1742-6596/241/i=1/a=012067 (Accessed October 19, 2012).

Egerton, R.F., Li, P. and Malac, M., 2004. Radiation damage in the TEM and SEM. *Micron, 35*(6): 399–409.

Findlay, S.D. and LeBeau, J.M., 2013. Detector non-uniformity in scanning transmission electron microscopy. *Ultramicroscopy, 124*: 52–60.

Fitting, L., Thiel, S., Schmehl, A., Mannhart, J. and Muller, D.A., 2006. Subtleties in ADF imaging and spatially resolved EELS: a case study of low-angle twist boundaries in SrTiO3. *Ultramicroscopy, 106*(11–12): 1053–1061.

Foord, R., Jones, R., Oliver, C.J. and Pike, E.R., 1969. The use of photomultiplier tubes for photon counting. *Applied optics, 8*(10): 1975–1989.

Forbes, B.D., d'Alfonso, A.J., Findlay, S.D., Van Dyck, D., LeBeau, J.M., Stemmer, S. and Allen, L.J., 2011. Thermal diffuse scattering in transmission electron microscopy. *Ultramicroscopy, 111*(12): 1670–1680.

Forbes, B.D., Martin, A.V., Findlay, S.D., D'Alfonso, A.J. and Allen, L.J., 2010. Quantum mechanical model for phonon excitation in electron diffraction and imaging using a born-oppenheimer approximation. *Physical Review B, 82*(10): 104103.

Fujinaka, S., Sato, Y., Teranishi, R. and Kaneko, K., 2020. Understanding of scanning-system distortions of atomic-scale scanning transmission electron microscopy images for accurate lattice parameter measurements. *Journal of Materials Science, 55*(19): 8123–8133.

Gnanasekaran, K., de With, G. and Friedrich, H., 2018. Quantification and optimization of ADF-STEM image contrast for beam-sensitive materials. *Royal Society Open Science, 5*(5): 171838.

Gonnissen, J., Batuk, D., Nataf, G.F., Jones, L., Abakumov, A.M., Van Aert, S., Schryvers, D. and Salje, E.K., 2016. Direct observation of ferroelectric domain walls in LiNbO3: wall-meanders, kinks, and local electric charges. *Advanced Functional Materials, 26*(42): 7599–7604.

Gonnissen, J., De Backer, A., Den Dekker, A.J., Martinez, G.T., Rosenauer, A., Sijbers, J. and Van Aert, S., 2014. Optimal experimental design for the detection of light atoms from high-resolution scanning transmission electron microscopy images. *Applied Physics Letters, 105*(6): 063116.

Grillo, V., 2011. An advanced study of the response of ADF detector. *Journal of Physics: Conference Series*, 326(1): 012036.

Grillo, V. and Rossi, F., 2011. A new insight on crystalline strain and defect features by STEM–ADF imaging. *Journal of Crystal Growth*, *318*(1): 1151–1156.

Hachtel, J.A., Idrobo, J.C. and Chi, M., 2018. Sub-Ångstrom electric field measurements on a universal detector in a scanning transmission electron microscope. *Advanced structural and chemical imaging*, *4*(1): 10.

Haruta, M., Fujiyoshi, Y., Nemoto, T., Ishizuka, A., Ishizuka, K. and Kurata, H., 2019. Extremely low count detection for EELS spectrum imaging by reducing CCD read-out noise. *Ultramicroscopy*, *207*: 112827.

House, S.D., Chen, Y., Jin, R. and Yang, J.C., 2017. High-throughput, semi-automated quantitative STEM mass measurement of supported metal nanoparticles using a conventional TEM/STEM. *Ultramicroscopy*, *182*: 145–155.

House, S.D., Schamp, C.T. and Yang, J.C., 2018. Real time acquisition and calibration of S/TEM probe current measurement simultaneously with any imaging or spectroscopic signal. *Microscopy and Microanalysis*, *24*(S1): 126–127.

Hovden, R. and Muller, D.A., 2012. Efficient elastic imaging of single atoms on ultrathin supports in a scanning transmission electron microscope. *Ultramicroscopy*, *123*: 59–65.

Isaacson, M.S., Langmore, J., Parker, N.W., Kopf, D. and Utlaut, M., 1976. The study of the adsorption and diffusion of heavy atoms on light element substrates by means of the atomic resolution STEM. *Ultramicroscopy*, *1*(3–4): 359–376.

Ishikawa, R., Lupini, A.R., Findlay, S.D. and Pennycook, S.J., 2014. Quantitative annular dark field electron microscopy using single electron signals. *Microscopy and Microanalysis*, *20*(1): 99–110.

Ishizuka, K., 2002. A practical approach for STEM image simulation based on the FFT multislice method. *Ultramicroscopy*, *90*(2–3): 71–83.

Jeong, J.S. and Mkhoyan, K.A., 2016. Improving signal-to-noise ratio in scanning transmission electron microscopy energy-dispersive X-ray (STEM-EDX) spectrum images using single-atomic-column cross-correlation averaging. *Microscopy and Microanalysis*, *22*(3): 536–543.

Jones, L., 2016. Quantitative ADF STEM: acquisition, analysis and interpretation. *IOP Conference Series: Materials Science and Engineering*, *109*(1): 012008.

Jones, L. and Downing, C., 2018. The MTF & DQE of annular dark field STEM: implications for low-dose imaging and compressed sensing. *Microscopy and Microanalysis*, *24*(S1): 478–479.

Jones, L. and Nellist, P.D., 2013. Identifying and correcting scan noise and drift in the scanning transmission electron microscope. *Microscopy and Microanalysis*, *19*(4): 1050.

Jones, L., MacArthur, K.E., Fauske, V.T., van Helvoort, A.T. and Nellist, P.D., 2014. Rapid estimation of catalyst nanoparticle morphology and atomic-coordination by high-resolution Z-contrast electron microscopy. *Nano Letters*, *14*(11): 6336–6341.

Jones, L., Varambhia, A., Beanland, R., Kepaptsoglou, D., Griffiths, I., Ishizuka, A., Azough, F., Freer, R., Ishizuka, K., Cherns, D. and Ramasse, Q.M., 2018a. Managing dose-, damage-and data-rates in multi-frame spectrum-imaging. *Microscopy*, *67*(suppl 1): i98–i113.

Jones, L., Varambhia, A., Sawada, H. and Nellist, P.D., 2018b. An optical configuration for fastidious STEM detector calibration and the effect of the objective-lens pre-field. *Journal of microscopy*, *270*(2): 176–187.

Jones, L., Yang, H., Pennycook, T.J., Marshall, M.S., Van Aert, S., Browning, N.D., Castell, M.R. and Nellist, P.D., 2015. Smart align—a new tool for robust non-rigid registration of scanning microscope data. *Advanced Structural and Chemical Imaging*, *1*(1): 1–16.

Kaneko, T., Saitow, A., Fujino, T., Okunishi, E. and Sawada, H., 2014. Development of a high-efficiency DF-STEM detector. *Journal of Physics: Conference Series*, *522*(1): 012050.

Katz-Boon, H., Rossouw, C.J., Dwyer, C. and Etheridge, J., 2013. Rapid measurement of nanoparticle thickness profiles. *Ultramicroscopy*, *124*: 61–70.

Katz-Boon, H., Walsh, M., Dwyer, C.J., Mulvaney, P., Funston, A.M. and Etheridge, J., 2015. Stability of crystal facets in Gold nanorods. *Nano Letters*, *15*: 1635–1641.

Keeney, L., Downing, C., Schmidt, M., Pemble, M.E., Nicolosi, V. and Whatmore, R.W., 2017. Direct atomic scale determination of magnetic ion partition in a room temperature multiferroic material. *Scientific Reports*, *7*(1): 1–11.

Kimoto, K. and Ishizuka, K., 2011. Spatially resolved diffractometry with atomic-column resolution. *Ultramicroscopy*, *111*(8): 1111–1116.

Kimoto, K., Asaka, T., Yu, X., Nagai, T., Matsui, Y. and Ishizuka, K., 2010. Local crystal structure analysis with several picometer precision using scanning transmission electron microscopy. *Ultramicroscopy*, *110*(7): 778–782.

Kimoto, K., Ishizuka, K., Asaka, T., Nagai, T. and Matsui, Y., 2005. 0.23 eV energy resolution obtained using a cold field-emission gun and a streak imaging technique. *Micron*, *36*(5): 465–469.

Kirkland, E.J., 1998. *Advanced Computing in Electron Microscopy*, Springer Science & Business Media.

Kirkland, E.J. and Thomas, M.G., 1996. A high efficiency annular dark field detector for STEM. *Ultramicroscopy*, *62*(1–2): 79–88.

Klenov, D.O. and Stemmer, S., 2006. Contributions to the contrast in experimental high-angle annular dark-field images. *Ultramicroscopy*, *106*(10): 889–901.

Koch, C.T., 2002. Determination of Core Structure Periodicity and Point Defect Density Along Dislocations. Dissertation:Ph. D., Arizona State University.

Kotaka, Y., 2010. Essential experimental parameters for quantitative structure analysis using spherical aberration-corrected HAADF-STEM. *Ultramicroscopy*, *110*(5): 555–562.

Kovarik, L., Stevens, A., Liyu, A. and Browning, N.D., 2016. Implementing an accurate and rapid sparse sampling approach for low-dose atomic resolution STEM imaging. *Applied Physics Letters*, *109*(16): 164102.

Krause, F.F., Schowalter, M., Grieb, T., Müller-Caspary, K., Mehrtens, T. and Rosenauer, A., 2016. Effects of instrument imperfections on quantitative scanning transmission electron microscopy. *Ultramicroscopy*, *161*: 146–160.

Krivanek, O.L., Chisholm, M.F., Nicolosi, V., Pennycook, T.J., Corbin, G.J., Dellby, N., Murfitt, M.F., Own, C.S., Szilagyi, Z.S., Oxley, M.P. and Pantelides, S.T., 2010. Atom-by-atom structural and chemical analysis by annular dark-field electron microscopy. *Nature*, *464*(7288): 571–574.

Lazar, S., Etheridge, J., Dwyer, C., Freitag, B. and Botton, G.A., 2011. Atomic resolution imaging using the real-space distribution of electrons scattered by a crystalline material. *Acta Crystallographica Section A: Foundations of Crystallography*, *67*(5): 487–490.

LeBeau, J.M. and Stemmer, S., 2008. Experimental quantification of annular dark-field images in scanning transmission electron microscopy. *Ultramicroscopy*, *108*(12): 1653–1658.

LeBeau, J.M., Findlay, S.D., Allen, L.J. and Stemmer, S., 2010. Standardless atom counting in scanning transmission electron microscopy. *Nano Letters*, *10*(11): 4405–4408.

Lefebvre, W., Hernandez-Maldonado, D., Moyon, F., Cuvilly, F., Vaudolon, C., Shinde, D. and Vurpillot, F., 2015. HAADF-STEM atom counting in atom probe tomography specimens: towards quantitative correlative microscopy. *Ultramicroscopy*, *159*: 403–412.

Li, X., Dyck, O., Kalinin, S.V. and Jesse, S., 2018. Compressed sensing of scanning transmission electron microscopy (STEM) with nonrectangular scans. *Microscopy and Microanalysis*, *24*(6): 623–633.

Lotnyk, A., Poppitz, D., Gerlach, J.W. and Rauschenbach, B., 2014. Direct imaging of light elements by annular dark-field aberration-corrected scanning transmission electron microscopy. *Applied Physics Letters*, *104*(7): 071908.

MacArthur, K.E., Brown, H.G., Findlay, S.D. and Allen, L.J., 2017. Probing the effect of electron channelling on atomic resolution energy dispersive X-ray quantification. *Ultramicroscopy*, *182*: 264–275.

MacArthur, K.E., D'Alfonso, A.J., Ozkaya, D., Allen, L.J. and Nellist, P.D., 2015. Optimal ADF STEM imaging parameters for tilt-robust image quantification. *Ultramicroscopy*, *156*: 1–8.

MacArthur, K.E., Jones, L.B. and Nellist, P.D., 2014. How flat is your detector? Non-uniform annular detector sensitivity in STEM quantification. *Journal of Physics: Conference Series*, *522*(1): 012018.

Maccagnano-Zacher, S.E., Mkhoyan, K.A., Kirkland, E.J. and Silcox, J., 2008. Effects of tilt on high-resolution ADF-STEM imaging. *Ultramicroscopy*, *108*(8): 718–726.

Martinez, G.T., De Backer, A., Rosenauer, A., Verbeeck, J. and Van Aert, S., 2014. The effect of probe inaccuracies on the quantitative model-based analysis of high angle annular dark field scanning transmission electron microscopy images. *Micron*, *63*: 57–63.

Martinez, G.T., Jones, L., De Backer, A., Béché, A., Verbeeck, J., Van Aert, S. and Nellist, P.D., 2015. Quantitative STEM normalisation: the importance of the electron flux. *Ultramicroscopy*, *159*: 46–58.

Mevenkamp, N., Binev, P., Dahmen, W., Voyles, P.M., Yankovich, A.B. and Berkels, B., 2015. Poisson noise removal from high-resolution STEM images based on periodic block matching. *Advanced Structural and Chemical Imaging*, *1*(1): 3.

Meyer, J.C., Kotakoski, J. and Mangler, C., 2014. Atomic structure from large-area, low-dose exposures of materials: a new route to circumvent radiation damage. *Ultramicroscopy*, *145*: 13–21.

Mittelberger, A., Kramberger, C. and Meyer, J.C., 2018. Software electron counting for low-dose scanning transmission electron microscopy. *Ultramicroscopy*, *188*: 1–7.

Mkhoyan, K.A., Maccagnano-Zacher, S.E., Kirkland, E.J. and Silcox, J., 2008. Effects of amorphous layers on ADF-STEM imaging. *Ultramicroscopy*, *108*(8): 791–803.

Muller, A. and Grazul, J., 2001. Optimizing the environment for sub-0.2 nm scanning transmission electron microscopy. *Journal of Electron Microscopy*, *50*(3): 219–226.

Muller, D.A., Kirkland, E.J., Thomas, M.G., Grazul, J.L., Fitting, L. and Weyland, M., 2006. Room design for high-performance electron microscopy. *Ultramicroscopy*, *106*(11–12): 1033–1040.

Müller-Caspary, K., Oppermann, O., Grieb, T., Krause, F.F., Rosenauer, A., Schowalter, M., Mehrtens, T., Beyer, A., Volz, K. and Potapov, P., 2016. Materials characterisation by angle-resolved scanning transmission electron microscopy. *Scientific Reports*, *6*: 37146.

Negi, D., Jones, L., Idrobo, J.C. and Rusz, J., 2018. Proposal for a three-dimensional magnetic measurement method with nanometer-scale depth resolution. *Physical Review B*, *98*(17): 174409.

Negi, D., Zeiger, P.M., Jones, L., Idrobo, J.C., van Aken, P.A. and Rusz, J., 2019. Prospect for detecting magnetism of a single impurity atom using electron magnetic chiral dichroism. *Physical Review B*, *100*(10): 104434.

Nellist, P.D., Lozano-Perez, S. and Ozkaya, D., 2010, January. Towards quantitative analysis of core-shell catalyst nano-particles by aberration corrected high angle annular dark field STEM and EDX. In *Journal of Physics: Conference Series, 241*: 012067

Nguyen, D.T., Findlay, S.D. and Etheridge, J., 2014. The spatial coherence function in scanning transmission electron microscopy and spectroscopy. *Ultramicroscopy*, *146*: 6–16.

Nord, M., Vullum, P.E., MacLaren, I., Tybell, T. and Holmestad, R., 2017. Atomap: a new software tool for the automated analysis of atomic resolution images using two-dimensional gaussian fitting. *Advanced Structural and Chemical Imaging*, *3*(1): 9.

Ooe, K., Seki, T., Ikuhara, Y. and Shibata, N., 2019. High contrast STEM imaging for light elements by an annular segmented detector. *Ultramicroscopy*, *202*: 148–155.

Ophus, C., 2019. Four-dimensional scanning transmission electron microscopy (4D-STEM): from scanning nanodiffraction to ptychography and beyond. *Microscopy and Microanalysis*, *25*(3): 563–582.

Ophus, C., Ciston, J. and Nelson, C.T., 2016. Correcting nonlinear drift distortion of scanning probe and scanning transmission electron microscopies from image pairs with orthogonal scan directions. *Ultramicroscopy*, *162*: 1–9.

Pennycook, T.J., Jones, L., Pettersson, H., Coelho, J., Canavan, M., Mendoza-Sanchez, B., Nicolosi, V. and Nellist, P.D., 2014. Atomic scale dynamics of a solid state chemical reaction directly determined by annular dark-field electron microscopy. *Scientific Reports*, *4*: 7555.

Pennycook, T.J., Lupini, A.R., Yang, H., Murfitt, M.F., Jones, L. and Nellist, P.D., 2015. Efficient phase contrast imaging in STEM using a pixelated detector. Part 1: experimental demonstration at atomic resolution. *Ultramicroscopy*, *151*: 160–167.

Pryor, A., Ophus, C. and Miao, J., 2017. A streaming multi-GPU implementation of image simulation algorithms for scanning transmission electron microscopy. *Advanced Structural and Chemical Imaging*, *3*(1): 15.

Rečnik, A., Möbus, G. and Šturm, S., 2005. IMAGE-WARP: a real-space restoration method for high-resolution STEM images using quantitative HRTEM analysis. *Ultramicroscopy*, *103*(4): 285–301.

Retsky, M., 1974. *Observed Single Atom Elastic Cross Sections in a Scanning Electron Microscope* (No. COO-1721-81). Dept. of Physics, Chicago Univ., Ill.(USA).

Rez, P., 2001. Scattering cross sections in electron microscopy and analysis. *Microscopy and Microanalysis*, *7*(4): 356–362.

Rez, P., 2003. Electron ionization cross sections for atomic subshells. *Microscopy and Microanalysis*, *9*(1): 42–53.

Rosenauer, A. and Schowalter, M. 2008. STEMSIM—a New Software Tool for Simulation of STEM HAADF Z-Contrast Imaging. Edited by A. G. Cullis and P. A. Midgley. *Microscopy of Semiconducting Materials 2007*, Springer Proceedings in Physics, 120.

Rosenauer, A., Mehrtens, T., Müller, K., Gries, K., Schowalter, M., Venkata Satyam, P., Bley, S., Tessarek, C., Hommel, D., Sebald, K., Seyfried, M., Gutowski, J., Avramescu, A., Engl, K., Lutgen, S., 2011. Composition mapping in InGaN by scanning transmission electron microscopy. *Ultramicroscopy* *111*(8): 1316–1327.

Rossouw, C.J., Dwyer, C., Katz-Boon, H. and Etheridge, J., 2014. Channelling contrast analysis of lattice images: conditions for probe-insensitive STEM. *Ultramicroscopy*, *136*: 216–223.

Ruiz-Marín, N., Reyes, D.F., Braza, V., Gonzalo, A., Ben, T., Flores, S., Utrilla, A.D., Ulloa, J.M. and González, D., 2019. Nitrogen mapping from ADF imaging analysis in quaternary dilute nitride superlattices. *Applied Surface Science*, *475*: 473–478.

Sanders, T. and Dwyer, C., 2018. Inpainting versus denoising for dose reduction in STEM. *Microscopy and Microanalysis*, *24*(S1): 482–483.

Sang, X. and LeBeau, J.M., 2014. Revolving scanning transmission electron microscopy: correcting sample drift distortion without prior knowledge. *Ultramicroscopy*, *138*: 28–35.

Sang, X. and LeBeau, J.M., 2016. Characterizing the response of a scintillator-based detector to single electrons. *Ultramicroscopy*, *161*: 3–9.

Sang, X., Grimley, E.D., Niu, C., Irving, D.L. and LeBeau, J.M., 2015. Direct observation of charge mediated lattice distortions in complex oxide solid solutions. *Applied Physics Letters*, *106*(6): 061913.

Sang, X., Lupini, A.R., Unocic, R.R., Chi, M., Borisevich, A.Y., Kalinin, S.V., Endeve, E., Archibald, R.K. and Jesse, S., 2016. Dynamic scan control in STEM: spiral scans. *Advanced Structural and Chemical Imaging*, *2*(1): 6.

Schwarzhuber, F., Melzl, P., Pöllath, S. and Zweck, J., 2018. Introducing a non-pixelated and fast centre of mass detector for differential phase contrast microscopy. *Ultramicroscopy*, *192*: 21–28.

Shibata, N., 2019. Atomic-resolution differential phase contrast electron microscopy. *Journal of the Ceramic Society of Japan*, *127*(10): 708–714.

Shibata, N., Findlay, S.D., Kohno, Y., Sawada, H., Kondo, Y. and Ikuhara, Y., 2012. Differential phase-contrast microscopy at atomic resolution. *Nature Physics*, *8*(8): 611–615.

Shibata, N., Kohno, Y., Findlay, S.D., Sawada, H., Kondo, Y. and Ikuhara, Y., 2010. New area detector for atomic-resolution scanning transmission electron microscopy. *Journal of Electron Microscopy*, *59*(6): 473–479.

Singhal, A., Yang, J.C. and Gibson, J.M., 1997. STEM-based mass spectroscopy of supported Re clusters. *Ultramicroscopy*, *67*(1–4): 191–206.

So, Y.G. and Kimoto, K., 2012. Effect of specimen misalignment on local structure analysis using annular dark-field imaging. *Journal of Electron Microscopy*, *61*(4): 207–215.

Song, B., Ding, Z., Allen, C.S., Sawada, H., Zhang, F., Pan, X., Warner, J., Kirkland, A.I. and Wang, P., 2018. Hollow electron ptychographic diffractive imaging. *Physical Review Letters*, *121*(14): 146101.

Spiegelberg, J., Idrobo, J.C., Herklotz, A., Ward, T.Z., Zhou, W. and Rusz, J., 2018. Local low rank denoising for enhanced atomic resolution imaging. *Ultramicroscopy*, *187*: 34–42.

Spurgeon, S.R., Balachandran, P.V., Kepaptsoglou, D.M., Damodaran, A.R., Karthik, J., Nejati, S., Jones, L., Ambaye, H., Lauter, V., Ramasse, Q.M. and Lau, K.K., 2015. Polarization screening-induced magnetic phase gradients at complex oxide interfaces. *Nature Communications*, *6*(1): 1–11.

Spurgeon, S.R., Du, Y. and Chambers, S.A., 2017. Measurement error in atomic-scale scanning transmission electron microscopy—energy-dispersive X-ray spectroscopy (STEM-EDS) mapping of a model oxide interface. *Microscopy and Microanalysis*, *23(3)*: 513–517.

Stevens, A., Yang, H., Carin, L., Arslan, I. and Browning, N.D., 2014. The potential for bayesian compressive sensing to significantly reduce electron dose in high-resolution STEM images. *Microscopy*, *63*(1): 41–51.

Stevens, A., Yang, H., Hao, W., Jones, L., Ophus, C., Nellist, P.D. and Browning, N.D., 2018. Subsampled STEM-ptychography. *Applied Physics Letters*, *113*(3): 033104.

Sun, Y. and Pang, J.H., 2006. AFM image reconstruction for deformation measurements by digital image correlation. *Nanotechnology*, *17*(4): 933.

Tate, M.W., Purohit, P., Chamberlain, D., Nguyen, K.X., Hovden, R., Chang, C.S., Deb, P., Turgut, E., Heron, J.T., Schlom, D.G. and Ralph, D.C., 2016. High dynamic range pixel array detector for scanning transmission electron microscopy. *Microscopy and Microanalysis*, *22*(1): 237–249.

Van Aert, S., Batenburg, K.J., Rossell, M.D., Erni, R. and Van Tendeloo, G., 2011. Three-dimensional atomic imaging of crystalline nanoparticles. *Nature*, *470*(7334): 374–377.

Van Aert, S., De Backer, A., Jones, L., Martinez, G.T., Béché, A. and Nellist, P.D., 2019. Control of knock-on damage for 3D atomic scale quantification of nanostructures: making every electron count in scanning transmission electron microscopy. *Physical Review Letters*, *122*(6): 066101.

Van Aert, S., De Backer, A., Martinez, G.T., Goris, B., Bals, S., Van Tendeloo, G. and Rosenauer, A., 2013. Procedure to count atoms with trustworthy single-atom sensitivity. *Physical Review B*, *87*(6): 064107.

Van den Broek, W., Reed, B.W., Béché, A., Velazco, A., Verbeeck, J. and Koch, C.T., 2019. Various compressed sensing setups evaluated against shannon sampling under constraint of constant illumination. *IEEE Transactions on Computational Imaging*, *5*(3): 502–514.

Van Dyck, D. and De Beeck, M.O., 1996. A simple intuitive theory for electron diffraction. *Ultramicroscopy*, *64*(1–4): 99–107.
Walther, T., 2006. A new experimental procedure to quantify annular dark field images in scanning transmission electron microscopy. *Journal of Microscopy*, *221*(2): 137–144.
Wang, J., Wang, J., Hou, Y. and Lu, Q., 2010. Self-manifestation and universal correction of image distortion in scanning tunneling microscopy with spiral scan. *Review of Scientific Instruments*, *81*(7): 073705.
Wang, Y., Suyolcu, Y.E., Salzberger, U., Hahn, K., Srot, V., Sigle, W. and van Aken, P.A., 2018. Correcting the linear and nonlinear distortions for atomically resolved STEM spectrum and diffraction imaging. *Microscopy*, *67*(suppl 1): i114–i122.
Wang, Z.W., Li, Z.Y., Park, S.J., Abdela, A., Tang, D. and Palmer, R.E., 2011. Quantitative Z-contrast imaging in the scanning transmission electron microscope with size-selected clusters. *Physical Review B*, *84*(7): 073408.
Wen, J., Mabon, J., Lei, C., Burdin, S., Sammann, E., Petrov, I., Shah, A.B., Chobpattana, V., Zhang, J., Ran, K. and Zuo, J.M., 2010. The formation and utility of sub-angstrom to nanometer-sized electron probes in the aberration-corrected transmission electron microscope at the university of Illinois. *Microscopy and Microanalysis*, *16*(2): 183.
Wenner, S., Jones, L., Marioara, C.D. and Holmestad, R., 2017. Atomic-resolution chemical mapping of ordered precipitates in Al alloys using energy-dispersive X-ray spectroscopy. *Micron*, *96*: 103–111.
Wu, X. and Baribeau, J.M., 2009. Composition and strain contrast of Si 1 − x Ge x (x = 0.20) and Si 1 − y C y (y ≤ 0.015) epitaxial strained films on (100) Si in annular dark field images. *Journal of Applied Physics*, *105*(4): 043517.
Yamashita, S., Koshiya, S., Ishizuka, K. and Kimoto, K., 2015. Quantitative annular dark-field imaging of single-layer graphene. *Microscopy*, *64*(2): 143–150.
Yankovich, A.B., Berkels, B., Dahmen, W., Binev, P., Sanchez, S.I., Bradley, S.A., Li, A., Szlufarska, I. and Voyles, P.M., 2014. Picometre-precision analysis of scanning transmission electron microscopy images of platinum nanocatalysts. *Nature Communications*, *5*(1): 1–7.
Yankovich, A.B., Zhang, C., Oh, A., Slater, T.J., Azough, F., Freer, R., Haigh, S.J., Willett, R. and Voyles, P.M., 2016. Non-rigid registration and non-local principle component analysis to improve electron microscopy spectrum images. *Nanotechnology*, *27*(36): 364001.
Yu, Z., Muller, D.A. and Silcox, J., 2004. Study of strain fields at a-Si/c-Si interface. *Journal of Applied Physics*, *95*(7): 3362–3371.
Zhang, Q., Zhang, L.Y., Jin, C.H., Wang, Y.M. and Lin, F., 2019. CalAtom: a software for quantitatively analysing atomic columns in a transmission electron microscope image. *Ultramicroscopy*, *202*: 114–120.
Zhu, Y. and Dwyer, C., 2014. Quantitative position-averaged k-, l-, and m-shell core-loss scattering in stem. *Microscopy and Microanalysis*, *20*: 1070–1077.
Zobelli, A., Woo, S.Y., Tararan, A., Tizei, L.H., Brun, N., Li, X., Stéphan, O., Kociak, M. and Tencé, M., 2019. Spatial and spectral dynamics in STEM hyperspectral imaging using random scan patterns. *Ultramicroscopy*, *212*: 112912.
Zuo, J.M., Shah, A.B., Kim, H., Meng, Y., Gao, W. and Rouviére, J.L., 2014. Lattice and strain analysis of atomic resolution Z-contrast images based on template matching. *Ultramicroscopy*, *136*: 50–60.

2 Machine Learning for Electron Microscopy

Alex Belianinov
Center for Nanophase Materials Sciences, Oak Ridge National Laboratory, Oak Ridge, TN 37831, USA

CONTENTS

2.1 Introduction 41
2.2 Machine Learning for Electron Microscopy 43
2.2.1 Supervised Learning in STEM 44
2.2.2 Dimensionality Reduction and Unsupervised Learning in STEM 49
2.2.3 Data Requirements 57
2.2.3.1 Present or Near Term 57
2.2.3.2 Data Lifecycle 58
2.2.3.3 Data-Centric Requirements: Capabilities, Speeds, and Feeds 61
2.2.3.4 Impediments, Gaps, Needs, Challenges 61
2.3 Outlook 62
Acknowledgments 63
References 63

2.1 INTRODUCTION

We are living a paradox. Global industrial and information progress has shaped a world that is simultaneously richer and more volatile than ever before (Treverton 2017). Rapid technological change will create economic and scientific opportunities while aggravating the divide between the haves and have nots. One of the forces in this tectonic technological shift is artificial intelligence (AI). Because innovation is a major source of economic growth (Drucker 1994; Oliner and Sichel 2002), penetration of AI into various aspects of science and technology is an inevitable reality likely to direct development of research problems, funding, instrumentation, and software.

Technology has grown at a breakneck pace, buttressed by Moore's law, which states that the speed and capability of computing devices can be expected to double every 2 years (Moore 1998). The sustainability of this rate is threatened in the long term. Although the innovation in design and semiconductor manufacturing maintains Moore's momentum, the classical cycle of synthesizing and commercializing new materials is lagging (Holdren 2011; Jain et al. 2013). To address the materials bottleneck, the material genome initiative (MGI) was designed to create a new era of policy, resources, and infrastructure that support U. S. institutions in the effort to discover, manufacture, and deploy advanced materials twice as fast, at a fraction of the cost (Green et al. 2017). The development of high-throughput experimentation for materials design will undoubtably be heralded as a major success for MGI (Hattrick-Simpers, Gregoire and Kusne 2016).

The next challenge in sustaining materials innovation is to design and implement a high-throughput characterization pipeline, akin to the MGI synthesis effort. Much of relevant material functionality is defined at scale inaccessible by first-principle simulations, operating within the constraint of current classical computational capabilities. There is hope this notion will be challenged by quantum computing supremacy, but this technology is still in the developmental stages. To speed up material development, efforts also must be directed to support analytical characterization methods. Here, the low-hanging fruit is in high-throughput methods capable of collecting physical and chemical information in a broad dynamic range of samples, scales, and conditions. Electron microscopies (EMs) fit the bill. These technologies have proven to be key to visualizing and analyzing matter at scales ranging from the micro to the atomic, offering a wealth of information on local structure, defects, and material dynamism (Gong et al. 2019). To bring EM to the necessary characterization, throughput levels will require a new research approach. This approach will have to rely on applying the emerging concepts in AI, machine vision, and deep learning and forge new paths in automating the extraction of relevant physics from this high-velocity volume of information.

In atomically resolved scanning transmission EM (STEM) images a priori symmetry assumptions have historically been the starting point of analysis. This fails with multiple phases and/or extended defects in images. A more clever approach is to find atoms without referencing a lattice, and many approaches are available, which will be discussed in some detail in the following sections. However, adaptation of a substantial portion of these algorithms is still nontrivial, especially with multiple chemical species in an image. Alternatively, images may be segmented to contain features of interest with relevant information extraction focusing only on these data subsets (Sarahan et al. 2011; Lu and Gauntt 2013). These contrast-based image analyses are powerful, and often one of the very few possible analysis approaches in two-dimensional (2D) images. It is important to remember that they too are prone to error propagation, especially with low-quality, noisy image data, and as such they may require rather extensive user involvement and expertise beyond microscopy. Efforts have been taken to simplify these processes and make them more robust via correcting algorithms with parameters of detector orientation and environmental influence (Jones and Nellist 2013). These image-processing techniques are powerful and have paved the way to maximize relevant data extraction, achieving impressive results that extend into three-dimensional (3D) reconstruction and internal structure mapping (Jones et al. 2014). Now, with even better AI-based methods and higher image quality, more ways are available to find, categorize, analyze, and interpret image parameters.

2.2 MACHINE LEARNING FOR ELECTRON MICROSCOPY

Arguably, an inescapable signpost to AI was the famous Turing Test, which hinges on the ability of machines to fool people into thinking that the machines are people (Marcus, Rossi and Veloso 2016). Partially autonomous and intelligent systems have been used in military technology since at least the World War II (Allen and Chan 2017), but advances in machine learning (ML) in the early 2000s produced the first widely used data mining and ML algorithms.

ML can be separated into two broad categories: supervised and unsupervised. In supervised learning the general task can be summarized as surmising a function that maps inputs to outputs based on a set of examples (Russell and Norvig 2016). In an ideal scenario, the supervised algorithm is capable of correctly labeling new data based on the class labels that were presented during training. A parallel task for humans is known as category learning, in which there is the search for and listing of attributes that can be used to distinguish exemplars from nonexemplars of various categories (Bruner, Goodnow and Austin 1956). Alternatively, in unsupervised learning, previously labeled examples do not exist, and algorithms must deduce previously unknown patterns in the data. These methods are commonly used for dimensionality reduction and clustering problems.

A popular example of supervised ML is the artificial neural network (ANN), a problem-solving architecture that mimics the brain, in which layers of neurons process signal via unique connectivity pathways and by adjusting synaptic weights. This creates an abstraction of data at each layer driving the decision-making process (Zurada 1992; Najem et al. 2018). Today, these types of networks are driving the development of graphical processing unit (GPU)-based neural network architecture, which offers significant improvement in performance (Oh and Jung 2004) and led to the subsequent development of deep neural networks (DNNs). The DNN approach allows an even higher level of feature abstraction and thus higher learning complexity (Goodfellow, Bengio and Courville 2016).

Many of these methods have been successfully applied in high-energy physics (Baldi, Sadowski and Whiteson 2014; Baldi et al. 2016; Guest et al. 2016) and biological (Angermueller et al. 2016) and medical imaging (Litjens et al. 2017). Some progress also has been made in material science (Strelcov et al. 2014; Belianinov, Kalinin and Jesse 2015; Belianinov et al. 2015; Jesse et al. 2016; Mueller, Kusne and Ramprasad 2016), but the effort has been fragmented. On the instrumentation front, the progress in STEM has allowed real-space imaging of structural and electronic material parameters on the atomic scale (Pennycook et al. 2008; Pennycook and Nellist 2011).The goal of imaging and spectroscopy in EM is to observe and quantify the structure and its identity with functionality by evaluating the breadth of physical chemical, electronic, optical, and phonon data of atoms and phases (Mody 2011). Better machinery and data-processing technologies allow for sub-10-pm positional precision (Kim et al. 2012; Yankovich et al. 2014), which enabled the visualization of chemical and mechanical strains (Kim et al. 2014) and ferroelectric polarization (Jia et al. 2007; Chang et al. 2011; Jia et al. 2011; Nelson et al. 2011) and octahedral tilts (Jia et al. 2009; Borisevich et al. 2010a,b; He et al. 2010; Kim et al. 2013). Now the technology is shifting focus away from hardware toward software and analytics.

This challenges the data-processing paradigms and poses new questions on how to better transfer, store, analyze, and learn from multidimensional datasets.

Even though the hardware is evolving, and much better software is leading to a rapid evolution, it is important to be reminded that there are immediate challenges rooted in the confidence of the data itself, before any analysis is performed. Some of the more relevant features in EM data that can aid with automation of identifying physically meaningful behavior include atomic positions (Belianinov et al. 2015), sample relevant crystallography and atomic neighborhoods (Ziatdinov et al. 2017), and feature groups (Vasudevan et al. 2015). All of these can be examined statically, or as they evolve with time (provided time series data were collected). For example, a significant advantage of Z-contrast STEM is that the contrast is associated with the position of atomic nuclei compared with the considerably more complex contrast formation mechanisms in transmission EM (TEM) (Kalinin et al. 2019). Thus, it is helpful to consider the level of confidence with which the atomic coordinates can be acquired, and how those are affected by the microscope itself. A more difficult question is the extent of generalization, from the imaged features on a single sample, or even area under the beam, to broader implications such as temperature and dopant/impurity distribution effects. These considerations aside, there is plenty of room for ML methods to be used in the analysis of scientific data for atom- and molecule-resolved EM experiments (Gómez-Bombarelli et al. 2018; Maksov et al. 2019).

2.2.1 Supervised Learning in STEM

A typical task common to many STEM image or movie analyses involves a conversion of images into a set of atomic coordinates or trajectories. There are several ways to achieve this, using less sophisticated tools, such as unsupervised approaches and image processing (LeBeau et al. 2008, 2010; Belianinov et al. 2015). Deep learning methods, however, succeed in working with data that cannot easily be reconciled by a human operator (Figure 2.1) (Kalinin et al. 2019).

For evolving image data, similar approaches may be used. Recently, a deep-learning framework for dynamic STEM imaging was demonstrated and used to locate defects in WS_2 (Maksov et al. 2019). The model allows highlighted defects from raw STEM data and classified them by type (Figure 2.2).

At the heart of the examples in Figures 2.1 and 2.2 is a deep convolutional neural network (DCNN) that infers the spatial pattern in the images via a grid-structured topology (Goodfellow, Bengio and Courville 2016). It is difficult to use this approach because the patterns learned are usually translationally invariant. The DCNN can be thought of as a set of operational units that can be broken down into sets of single convolutional layers consisting of independent weight matrices (filters). These operate by sliding across the image and taking a dot product with the area they cover. These vectors of dot product values are reassembled back into a 3D stack that is the same size as the input, with the depth defined by the number of weight matrices in the convolution layer. At each layer, parameters such as the size of the weight matrix and the sliding step side also need to be defined. The 3D output of each convolutional layer is passed through a digital rectifier, an activation function (Glorot, Bordes and Bengio 2011), which is also known as a rectified linear unit (ReLU)

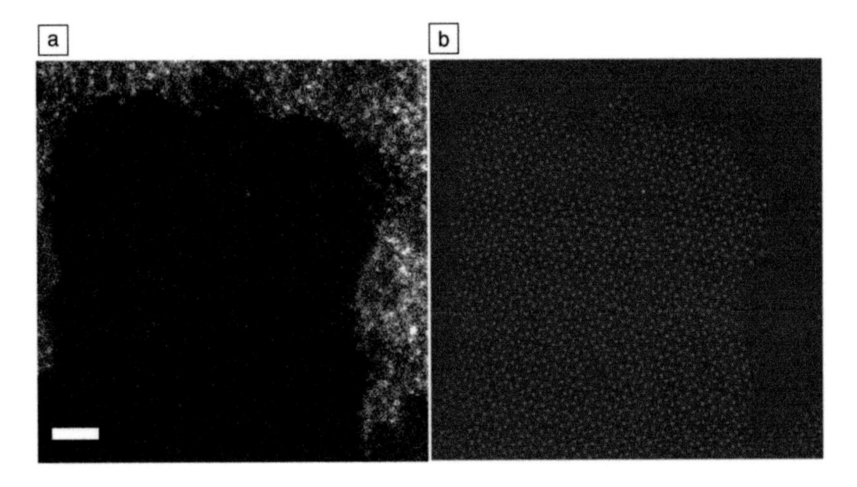

FIGURE 2.1 Application of deep-learning networks for feature finding and image inversion. (a) STEM image of Si adatoms on graphene in the presence of contaminants. Scale bar = 1 nm. These data show graphene lattice (darker regions with periodic pattern), amorphous SiC regions (larger brighter regions), and point Si dopants in graphene lattice (isolated bright spots). (b) deep-learning analytics of the image in (a). Green corresponds to Si atoms and red to carbon. The output represents the probability density that a certain pixel belongs to a particular atom type. Note the robustness of the network to noise when the locations of carbon atoms are identified above the human eye perception.

Reproduced with permission from (Kalinin et al. 2019).

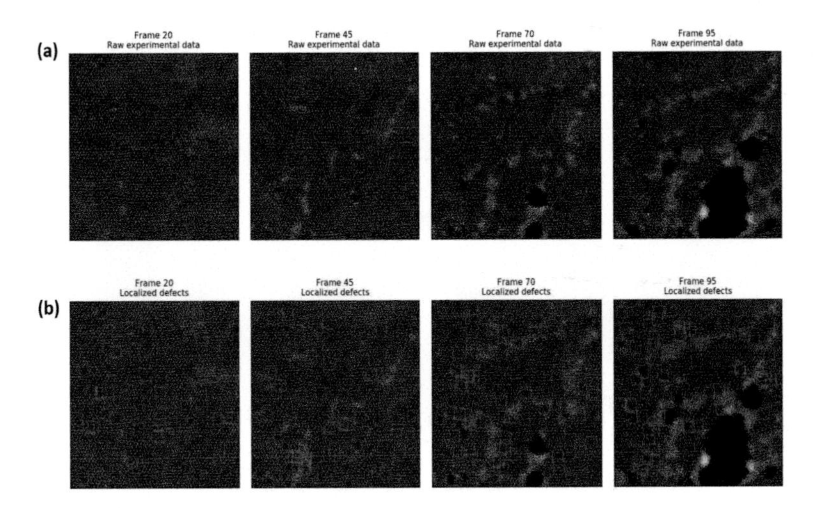

FIGURE 2.2 Defect evolution under e-beam irradiation in mo-doped WS_2. (a) Four selected frames from the STEM movie of mo-doped WS_2 obtained at 100 kV illustrating formation of defects and lattice transformations as a function of time. (b) Same data with *point-like defects* localized. Note that here we do not consider extended defects (Maksov et al. 2019).

Reproduced with permission from (Maksov et al. 2019).

(Nair and Hinton 2010). This is used to separate specific excitation and unspecific inhibition at each layer. This output is then passed on to the next convolutional layer. This allows the construction of a neural pyramid in which each layer may learn its own set of patterns. At the apex are local features that extend to more complex, long-range interactions. To make the process less error prone, max-pooling layers are also placed between convolutional layers to reduce the number of feature-map coefficients and to spike the likelihood of association in the spatial hierarchy by focusing the next convolutional operation on a larger region. This process is more intuitively illustrated in Figure 2.3.

Figure 2.3 Pixel-level labeling process in which the network is trained to identify the type and the location of an atom or a defect from an image. The encoder was designed as a set of feature extraction convolutional layers followed by max-pooling layers for reducing the data size (Ziatdinov et al. 2017). Convolutional layer filters doubled in number after each max-pooling layer. The input to this DCNN is a gray-scale image, and the output is a 3D tensor the size of the original image *x* by original image *y* by number of channels. A channel is defined as a circular structure and corresponds to a specific atom type such as nitrogen, silicon, and so forth. The center of the circle is the atomic location, or the (*x*, *y*) coordinate of the atomic position. Note that this approach requires training examples of the atoms or defects. These can be made via simulation or manual data labeling. The deconvolutional part of the network converted low-resolution feature maps into full input size maps, which were passed to a SoftMax layer for the pixel-wise classification of the probability of a pixel being an atom, or defect.

An excellent review of some of these methods (and many others) has been offered recently by Dan, Zhao and Pennycook (2019). The three main areas of confusion in which EM entrants into AI often find themselves are how to exactly to get started,

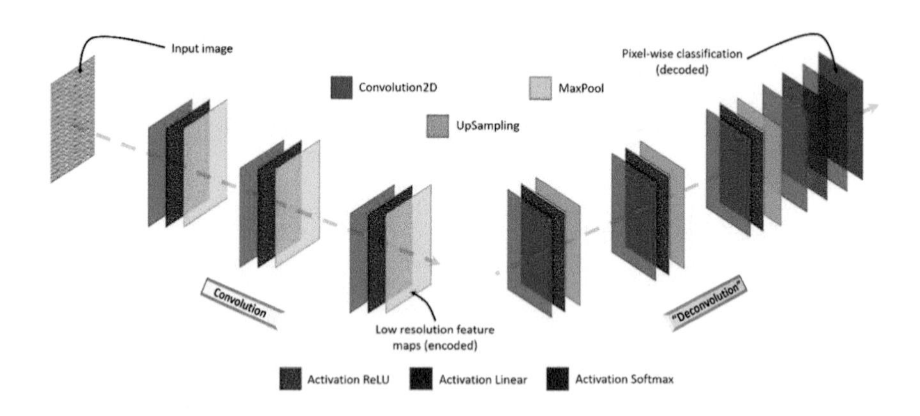

FIGURE 2.3 Application of deep learning to a problem of finding atomic species and defects in a crystal lattice. Schematic architecture of a fully convolutional network that has an encoder-decoder type of structure (or convolution-deconvolution structure). The final SoftMax layer outputs a pixel-wise classification for atomic species and/or defects.

Reprinted with permission from (Ziatdinov et al. 2017).

what model to use, and what the results mean. It seems unreasonable to attempt to understand what any model output means or may imply without relevant information on the specific data and the analysis used. However, some guidance in how to get started and how to go about building a model can be offered in the remainder of this section.

First, it is important to realize that many people in the field have done the heavy lifting and open source packages such as pyCroscopy, pyUSID, AICrystallographer, py4DSTEM, and DeepDiffraction are readily available on GitHub to get the absolute beginners going. These are well documented and heavily used, as such they are an excellent place to start for anyone new to programming. Please note that even these high-quality resources require the users to know the basics of programming environments, packages, language specifics, and file inputs/outputs to just get started. Unfortunately, many users that get to the initial levels of programing proficiency still get put off by the seeming complexity of building and running a neural network. Luckily, it is easy to do so for image processing and atom finding.

Let us consider the process of constructing a fully convolutional neural network. In its basic form, it will take a noisy image as an input and return atomic positions as an output. First, this network will need to be trained. There are many ways to make or acquire training data; one approach is to use density functional theory (DFT) to simulate a surface most relevant to your problem, obtain true atomic coordinates from the calculation, and then add noise for training. Alternatively, a lattice of 2D Gaussians can be created to represent atoms at known locations with specific types of noise later added to train the network. Regardless, a set of real or synthetic data in which the ground truth is known is necessary to get started.

Once the data are ready you may want to think about what library to use to create a network. For Python, the typical choices are Keras, Tensorflow, or Pytorch. Each has their own distinct strengths and weaknesses that are beyond the scope of this discussion. To create the network outlined in Figure 2.3, several blocks are required: a convolutional layer, a parametric ReLU activation layer, and a MaxPool layer for convolution; a convolution, ReLU activation, and UpSampling; and a pixel classifier for deconvolution. Additional parameters, such as the number of filters, padding, the size of the kernel, the step size, and dilation, must be defined to segment the data at each layer. This is by far the most cumbersome step, as it simultaneously forces the user to encode the structure of the network as well as define all the sizes, sampling, and data flow. After this step the model is compiled, and the network is ready to be trained. Remember, you will need a training pair, the training image, and the ground truth that you have collected or created. Depending on the size of your network and the amount of data you have, this may take some time. Validating your model, with data that your network has not been trained for, should be the last step before running data on interest.

A keen interest in deep learning on image data is to create a framework capable of making the transition to bridge the gap between experiment and modeling via ML. Li et al. (2017) demonstrated the use of ML and phase field modeling to quantify flexoelectricity in $PbTiO_3/SrTiO_3$ superlattice polar vortices. As a word of caution, using AI to run analysis by comparison between theory and experiment can be misleading. Specifically, comparing simulation or theory images output with the experimental data offers a qualitative analysis only, even though it may seem tempting to use a numeric similarity measure and thus claim to confirm results. In the same vein, small outliers in the images (or spectra) that do not affect the overall similarity measure may in fact be the very reason behind a given behavior of a material system. Li et al. (2017) predicted the polarization vortex structure for a number of flexoelectric coefficients, and then they utilized a structural similarity index to derive the flexoelectric coefficients of their material, (Figure 2.4).

Many of the ML approaches to STEM promise exciting possibilities for defect identification and propagation. Many of the algorithms are robust and can work with large data volumes. However, few challenges need to be addressed for mass acceptance and implementation of many of these approaches. The lack of standardization is a major turn off from using ML in STEM imaging processing. Because no single general training dataset exists, many ML models prove to be self-consistent only on a small subset of images. This also would aid in quantifying performance metrics of various approaches and act as a way to validate and check accuracy. Finally, many of the supervised techniques are more difficult to run and implement conceptually and without some specialized training. This may explain the overall popularity of the unsupervised methods discussed in the next section.

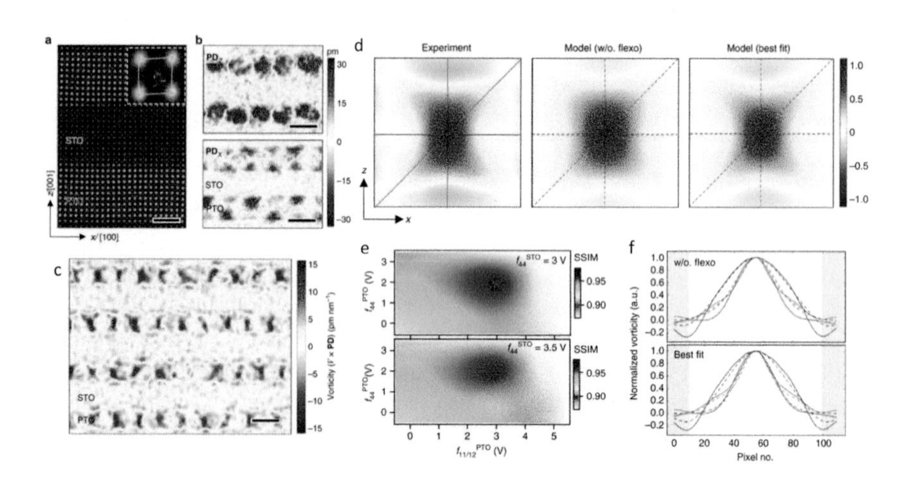

FIGURE 2.4 Computer vision methods for fitting models to imaging data. (a) STEM image of $PbTiO_3/SrTiO_3$ superlattice. (b) Derived distortion map along the *z*- and *x*-directions. (c) Derived vorticity map clearly showing the alternating vortex structures in the superlattice. (d) First principle component map of vortex structures from experiment, as well as from the simulations with and without the flexoelectric coupling term. (e) Structural similarity index as a function of the flexoelectric coupling coefficients, with the best fit marked by a star. (f) Line profile of the vorticity along the *x*-, *z*-, and *x-z* directions, with experiment results plotted as sold lines and simulations as dashed lines. Note the colors of the lines follow the scheme in (a). From Li et al. Nature communications (Li et al. 2017).

Reprinted with permission from reference (Li et al. 2017).

2.2.2 Dimensionality Reduction and Unsupervised Learning in STEM

Dimensionality reduction is used to reduce the number of random variables into a set of principal variables (Roweis and Saul 2000). There are many approaches to dimensionality reduction such as principal component analysis (PCA), matrix factorization, independent component analysis (ICA), and so forth (Hyvärinen and Oja 2000; Singh and Gordon 2008; He et al. 2015). These methods offer different levels of effectiveness for a given problem, and suffer from lack of interpretability in some cases. Nonetheless, due to their relative ease of implementation (usually much easier than supervised learning) and flexibility in data that can be used as input, these are great tools for exploratory image and spectral analysis in STEM.

Unsupervised learning is a type of ML algorithm used to draw inferences from datasets consisting of input data without labeled responses. Typically, these refer to clustering algorithms used to group data based on its similarity. The idea is to use a similarity metric that will allow the algorithm to distinguish data based on its own internal structure without any other assumptions. Cluster analysis includes a myriad of algorithms with perhaps the most common being k-means and Density-based spatial clustering of applications with noise (DBSCAN) (Arthur and Vassilvitskii 2007; Schubert et al. 2017).

PCA is a very well-known dimension reduction method. This linear transformation method finds the principal axes in data that maximize variance. As such, the first few principal components may act as a snapshot characteristic that describe the data. On the downside, the necessary condition of orthogonality for all principal component vectors may result in negative values in components, rendering aspects of the data as uninterpretable. However, this can be ameliorated by enforcing nonnegativity in the algorithm, or using another approach, namely nonnegative matrix factorization (NMF).

Figure 2.5 illustrates how PCA was used on the 4D STEM Ronchigram data to explore salient image features in a specific example of $BiFeO_3$ (BFO) domains (Jesse

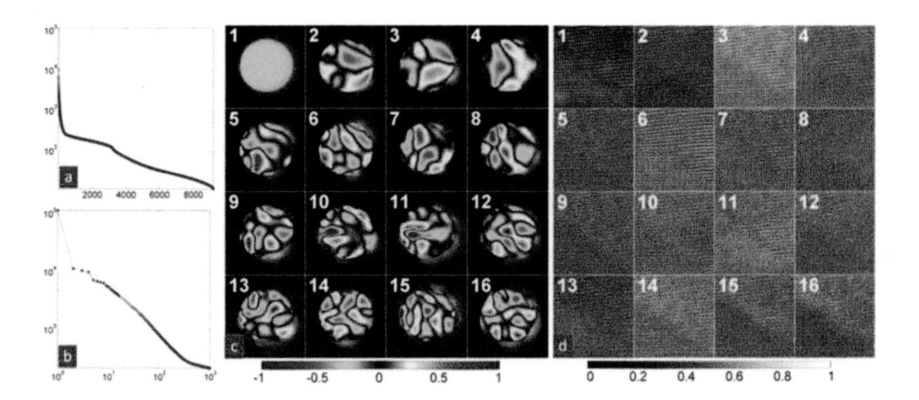

FIGURE 2.5 (a) Full log-linear scree plot of information content for all (96×96) 9216 principal components. (b) Log-log scree plot of the first 1000 principal components with the first 16 corresponding to images in (b) and (c) highlighted. (c) The first 16 PCA eigenvectors. (d) The first 16 PCA loading maps.

Reprinted with permission from (Jesse et al. 2016).

et al. 2016). Here, the use of PCA was an excellent denoising and a compression method that paved the way to a subsequent clustering analysis that was capable of distinguishing ferroelectric domains. Note that in these examples the authors are not explicitly trying to find the atoms and deduce the phase composition from atomic positions. Rather, relying on the simultaneously obtained information-rich diffraction data, the boundary reveals itself in the components (3, 15, 16). It is true that the principal Ronchigrams are likely nonsensical, or at least exceedingly difficult to interpret. This is not the point. In this analysis the Ronchigram data were the means to an end, which was ferroelectric domain analysis in BFO.

In NMF (Lee and Seung 1999) the input data are decomposed into nonnegative factors. NMF is popular among scientists for spatially resolved spectral analysis, defined as finding $k \ll m$ basic spectra (basis functions that change gradually with composition, in terms of structure and intensity), such that all the m measurements can be explained as a mixture of the k basic spectra (Kannan et al. 2018). Figure 2.6 illustrates the NMF decomposition of an atomically resolved Mo–V–Te–Nb oxide catalyst image after a sliding window fast Fourier transform (FFT) analysis (Vasudevan et al. 2015; Kannan et al. 2018).

The NMF analyses of the sliding FFT data are shown in Figure 2.6 b–g. Here, the chemical phases are shown as the first and second components in Figure 2.6b and e and c and f. The third component is the interface between phases. In this scenario NMF decomposition works based on diffraction by linearly deconvoluting spectra that sums to one. In addition, the algorithm acts to segment the image data based on phase.

These techniques are valuable to reveal the general data behavior through a low-rank approximation. Typically, these methods are easy to implement, are robust, and work well to at least filter noise from high-dimensionality data. The interpretability varies, especially in the dimensionality reduction set of algorithms. PCA and ICA suffer from negative entries, whereas NMFs rectify this deficiency. The most useful

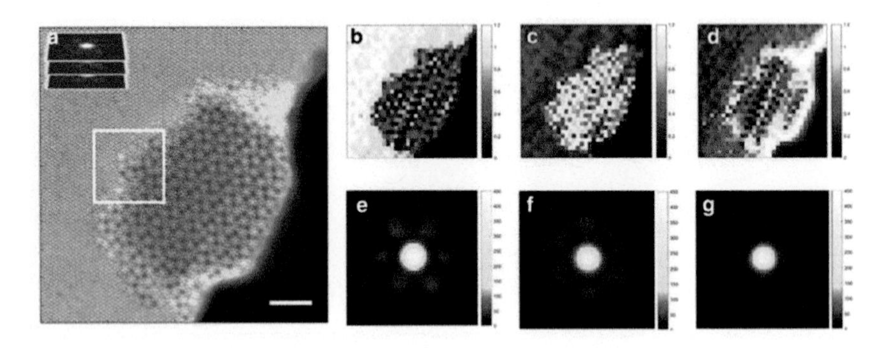

FIGURE 2.6 (a) Experimental STEM image of a Mo–V–Te–Nb oxide catalyst. The image size is 2048 × 2048 pixels, the width of the window in a sliding FFT is set to 500 pixels (shown schematically in the figure), and the window step size is 100 pixels. The top left corner inset shows schematically a stack of 2D FFT images formed during the sliding FFT procedure. Scale bar = 5 nm. (b–g) Results of NMF-based decomposition of sliding FFT data over the area into three components. Loading maps (b–d) associated with end members (e–g).

feature of these methods is that each component is representative of a region with similar behavior.

A popular clustering algorithm, k-means clustering, aims at assembling feature vectors into *k* groups with equal variance. The caveat with this technique is that the *k* value, or the number of clusters, must be user supplied. In some cases, this information can easily obtained, such as in the case of feature vectors describing the atomic structure, or a distribution of phases. Assigning the *k* gets more difficult as the data gets more dissimilar and noise starts playing a larger role.

In these cases, DBSCAN and other hierarchical clustering methods may be a better choice. DBSCAN is more apt at working with data that has features clustering in aggregates of different size and shape. The algorithm operates by estimating the density of the feature vectors via an input parameter ε, which represents the distance between feature vectors in the same cluster. Unfortunately, only a single ε can be specified in DBSCAN. The OPTICS algorithm can be a solution for cases when ε is a limitation. An additional feature for DBSCAN users interested in processing large data quantities is the parallelism offered by HDBSCAN. The main problem with the density-based clustering approach is that a small percentage of feature vectors may end up not labeled, complicating cluster structure. In these cases, a tool such as a dendrogram, or a clustering tree, can be helpful to pinpoint similarities among feature vectors and feature vector groups.

In Figure 2.7, the k-means analysis is performed on the atomically resolved images. All atoms are first found, and then a local neighborhood consisting of the

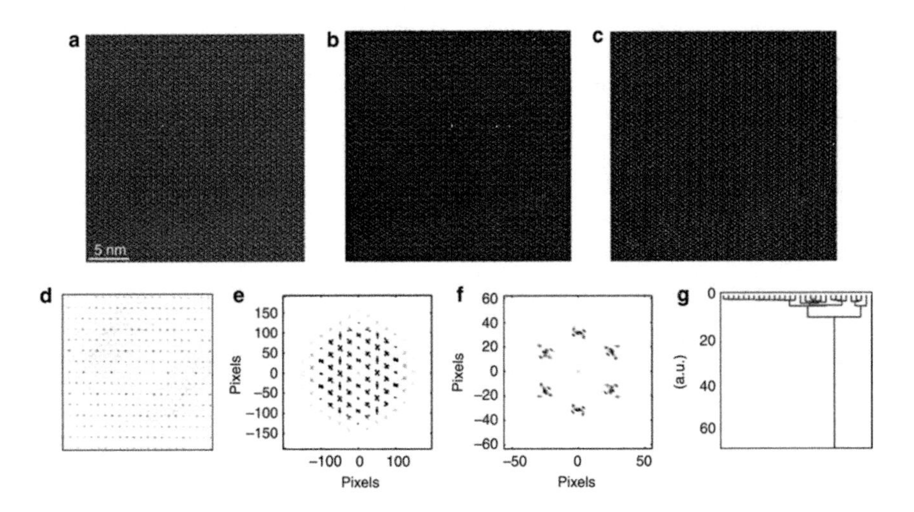

FIGURE 2.7 Single M2 phase Mo–V–Te oxide. (A) M2 phase STEM image. (b) The k-means clustering results for six neighbors, sorted by distance metric. (c) The k-means clustering results for six neighbors, sorted by angle metric. (d) FFT of the image in (a). (e) The 50-member neighborhood of the image in (a). (f) The six-member neighborhood of the image in (a). (g) Dendrogram for the six neighbors, sorted by angles metric, with the *y*-axis signifying the cluster separation in the hierarchical tree.

Reproduced with permission from (Belianinov et al, 2015).

nearest six atoms is constructed for each atom found. The lengths and angles to each of the neighbors is the feature vector for the k-means algorithm.

Figure 2.7 illustrates minute deviations of the internal M2 phase structure of Mo–V–Te oxide. The Fourier transform and the nearest neighbor histogram for 50 and 6 neighbors are shown in Figure 2.7d–f. Note that the FFT reflects the high degree of crystallinity of this area on the sample. In Figure 2.7e and f the histogram illustrates the internal neighborhood structure that is *x*, linear, and shaped like a dot. This is an indicator that the M2 phase can be considered as a hexagonal structure with several periodic distortions. The dot-like spots show the unit cell for the superstructure, whereas the distorted spots are the symmetry of the specific cells. The length- and angle-based k-means clustered image overlays are shown in Figure 2.7b and c. The antiphase boundary not evident from the raw data in Figure 2.7a becomes clear after the analysis. This local neighborhood analysis of materials from atomically resolved image data is capable of linking chemistry and physics at the atomic scale. The local bond characteristics such as the outlier in the coordination sphere as well as the bonding type can easily highlight the uniform phase regions, as well as mixed regions and defects.

There is another clustering approach used to detect, identify, and classify local structure in atomically resolved STEM images (Laanait et al. 2016). Here, scale invariance and contextual structural identification were utilized on synthetic and real data based on atomic column intensity distribution. Interestingly, the authors have shown that a more primitive structural state is sufficient to extract the atom positions.

The classification of columns of the $SrTiO_3$ and $SrTiO_3/BaTiO_3$ images via agglomerative clustering with different amounts of noise in the system are illustrated in Figure 2.8. Here, each class is shown by a different color. In bulk $SrTiO_3$, the process easily differentiates between different species in the unit cell. Interestingly, in the analysis the O1 and the O2 oxygen columns were grouped into different clusters, due to the descriptors being rotationally variant. Furthermore, when the noise was high the Sr and the Ti were still correctly identified. In $SrTiO_3/BaTiO_3$, full classification was possible with Ti columns in bulk STO, at the interface, and in bulk BTO. The results were comparable for Sr and Ba as well.

Note that the previous examples utilized a feature vector approach, where the vectors were designed with an end in mind. This approach helps to compress and encode relevant image information on texture, shape, distance, and so forth, into a descriptor to be operated on by an algorithm. Due to the difficulty of translating image aspects into algorithm actionable data, descriptors can be utilized to encode properties and offer greater flexibility in describing structure.

The following example presents the analysis of Ronchigram datasets with fuzzy topological sets using the Riemannian metric (Li et al. 2019) with a recently introduced uniform manifold approximation and projection (UMAP) approach. (McInnes, Healy and Melville 2018). UMAP builds on LargeVis (Tang et al. 2016) by building a nearest neighbor cluster tree using random projection (Dasgupta and Freund 2008) and then finding for the lower dimensional manifold by probabilistic and negative edge sampling (Dong, Moses and Li 2011; Mikolov et al. 2013). For clarification, the manifold structures can then be analyzed by various clustering approaches. The data analyzed by Li et al. (2019) consisted of a synthetic and an experimental dataset

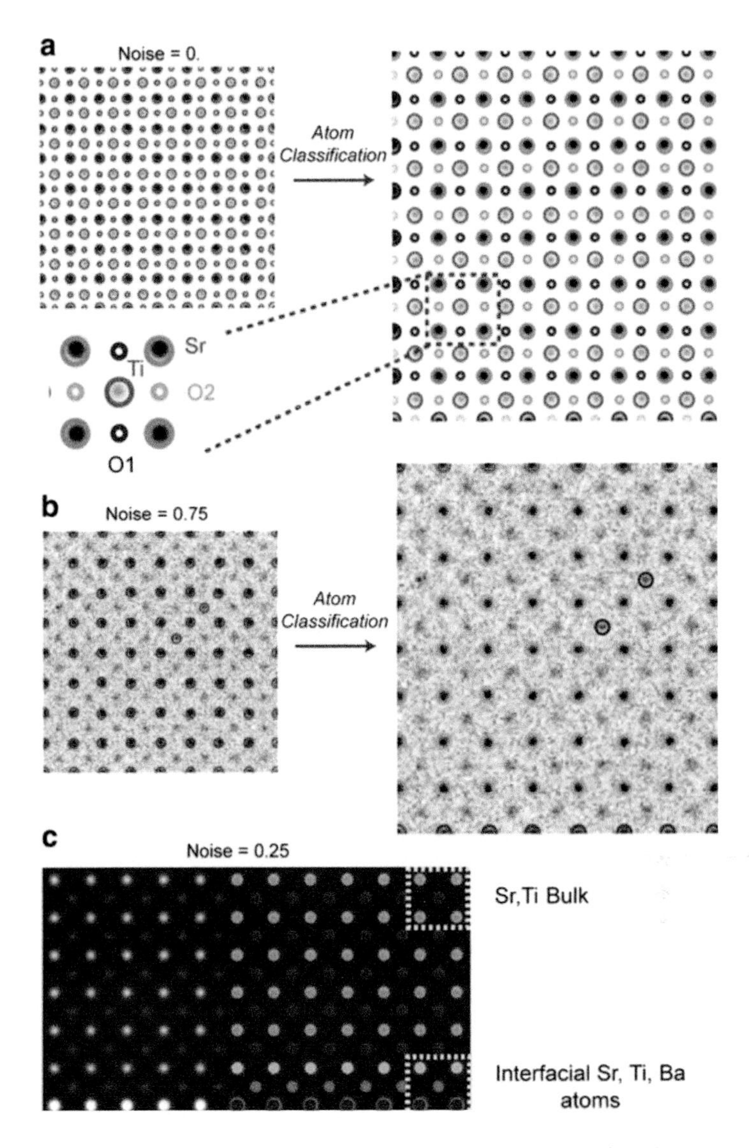

FIGURE 2.8 Classification of local structural states. Detected atomic columns at different noise levels in simulated STEM images are classified by hierarchical clustering, with different structural classes represented by circles with different colors. The different atomic columns in the [100] projection of the $SrTiO_3$ unit cell are all classified as distinct structural states by the presented approach (a). In the presence of noise, the distinction between Sr and Ti atomic columns is still maintained (b). Note that Sr atoms at the edge of the image belong to separate classes because their coordination is different than that of Sr atoms in the bulk. (c) Classification of atoms in the image of a $SrTiO_3$/$BaTiO_3$ interface distinguishes the interfacial atoms (Sr, Ti, Ba) from those present in the bulk phases, and it provides a complete description of the structural configurations present in the image.

Reproduced with permission from (Laanait et al. 2016)

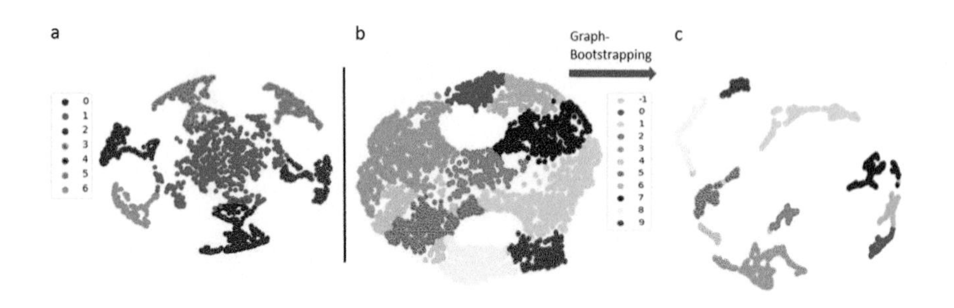

FIGURE 2.9 Manifold bootstrapping and clustering. (a) spectral clustering results on a UMAP manifold derived from the synthetic dataset. (b) UMAP manifold and (c) bootstrapped UMAP manifold derived from the experimental dataset. UMAP and bootstrapped UMAP manifolds are colored by the same set of clustering labels where HDBSCAN clustering was performed on the bootstrapped manifold. Bootstrapped manifold in (c) is derived from the reconstructed graph based on the UMAP manifold in (b).

Reproduced with permission from Li et al. (2019).

in which the simulated data size was (37, 64, 120, 140) with 37×64 electron probe positions and individual Ronchigrams of 120×140 pixels; the experimental data had 64×64 probe positions and 180×180 Ronchigram pixels sized at (64, 64, 180, 180). The Euclidean distance was used, with local neighborhood size (50) and the effective minimum distance between embedded points (o) as other tuned UMAP parameters.

Figure 2.9a illustrates UMAP Ronchigram hexagon projections with seven spectral clusters. Figure 2.9b is an experimental MAP manifold. The number of clusters was estimated on local structure parameters, such as the structure of the nearest neighbor and density. Figure 2.9c is a recalculated manifold using the hierarchical density estimate methods Hierarchical Density-Based Spatial Clustering of Applications with Noise [(HDBSCAN); Campello et al. 2015; McInnes, Healy and Astels 2017] to perform clustering on the bootstrapped UMAP manifold. The results of this approach on the synthetic data are shown in Figures 2.10 and 2.11.

Figure 2.10a illustrated the high-angle annular dark-field (HAADF) representation of the synthetic data. Figure 2.10b depicts the ground-truth atom positions over the spatial mapping of cluster labels derived from the manifold space shown in Figure 2.9a. The key detail here is that clusters 0, 2, 5 and 3, 4, 6 form two supergroups representing the graphene sublattices, and cluster 1 is the interatomic space. Figure 2.10c is the mean and standard deviation of Ronchigrams for each cluster with respect to the atomic positions.

Figure 2.11 is the similarity loadings of the seven assigned clusters. For every cluster, the distribution of bright blobs in the similarity loading is consistent with cluster label positions. Note that all loadings consistently show a black void around the fourfold Si dopant position in the center. This captures the change in the deflection patterns around the dopant.

Finally, Figure 2.12 shows the experimental data similarity cluster loadings. The clusters were also split into two groups A (clusters 0, 1, 5, 6) and B (clusters 2, 3, 7, 8). The similarity loadings also display a black void around the fourfold Si dopant

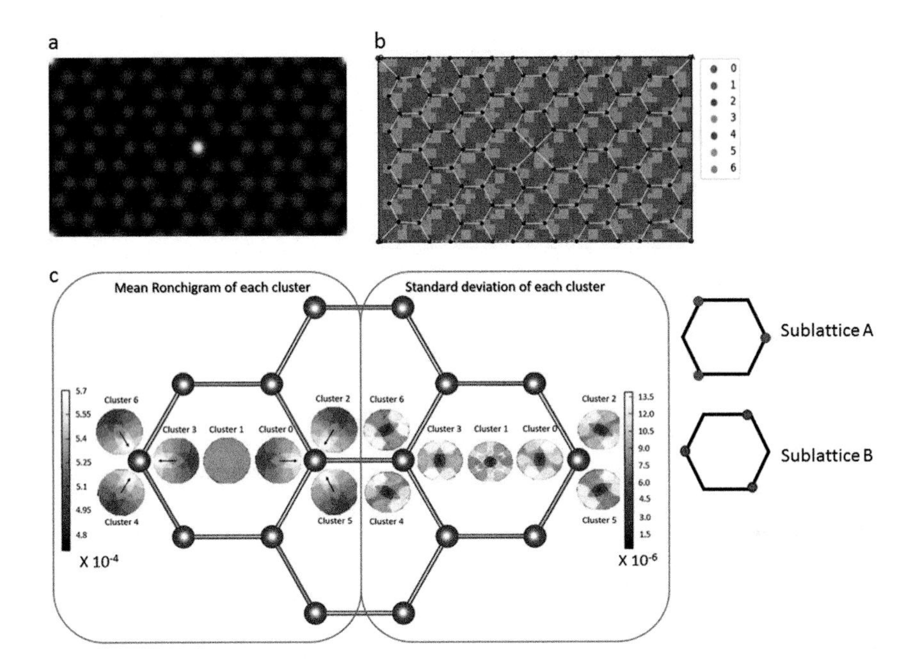

FIGURE 2.10 Synthetic data analysis. (a) High-angle annular dark-field image of the synthetic dataset. (b) Ground-truth atom positions overlaid on real-space distributions of cluster labels in Figure 2.2a. Black atoms are carbon, and blue atoms are the Si dopants. (c) The mean and standard deviation of Ronchigrams for each cluster and positions to the atom sites. Arrows indicate the simulated beam deflection observed near an atom. The Ronchigram intensity moves toward the atom.

Reproduced with permission from (Li et al. 2019).

position, suggesting a divergence between the Ronchigrams around the Si site, as opposed to the C site.

The novelty of this complex approach to solve a seemingly easy problem of finding a dopant atom lies in effectively trading off the spatial dimension for a potentially more information-rich diffraction space. Specifically, this approach is relevant for fruitful analysis of low-dose imaging data with the penultimate goal of atomic manipulation that currently relies on sequential imaging and probe positioning (Susi, Meyer and Kotakoski 2017; Dyck et al. 2018b).

In these discussions the question of whether a supervised or an unsupervised method is more appropriate will inevitably arise. Although there is no right or wrong, certain data streams are more appropriate for supervised learning while others are less so. The supervised methods are generally more difficult to set up and train; they also need a source of training data. However, once the infrastructure is established and the algorithms are trained, they work very quickly. It then seems that a supervised method is more appropriate for a type of problem that needs to be solved multiple times. Large datasets, or data that get continuously collected (many iterations of a similar sample type), are worthwhile to investigate

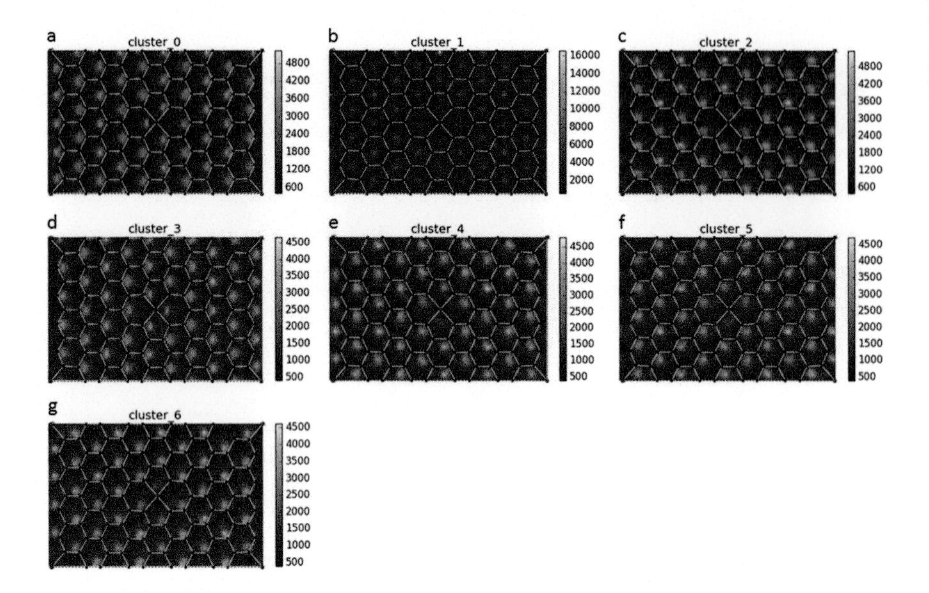

FIGURE 2.11 Similarity loadings of clusters from the synthetic dataset. (a, c, f) Similarity loadings of clusters 0, 2, 5 over sublattice A. (b) Similarity loadings of cluster 1 located in the space between atoms. (d, e, g) Similarity loadings of clusters 3, 4, 6 over sublattice B.

Reproduced with permission from Li et al. (2019).

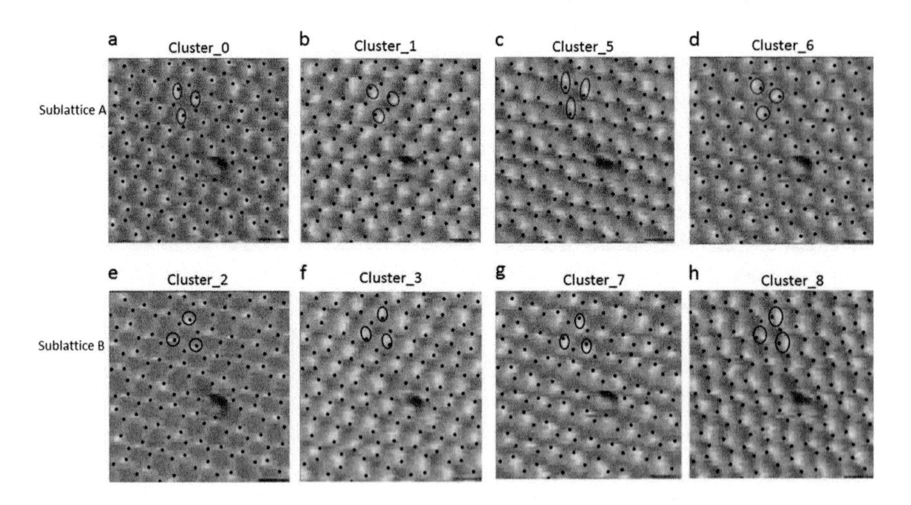

FIGURE 2.12 Similarity loadings of clusters from the experimental dataset. (a–d) Similarity loadings of clusters over sublattice A. (e–h) Similarity loadings of clusters over sublattice B. Here, we overlay the circle markers to illustrate the relative positions between atom sites and bright blobs.

Reproduced with permission from Li et al. (2019).

with supervised methods because the same network can be reused again without the training penalty. In these situations, the training data will likely come from experiments, which can be an advantage. Unsupervised methods are better for smaller problems, or problems that may need to be solved once. Once the data are wrangled into a suitable data structure, multiple compression, deconvolution, and clustering methods may be attempted and cross-validated. Some well-established methods are fast and parallelizable. Naturally, hybrid models in which some unsupervised methods inform the supervised approach, or a variation on this theme, are also possible. A word of caution, however. It is wise to keep reminding oneself of the analysis aim. Many hours can be spent running and optimizing these tools to hit an arbitrary benchmark without ever getting closer to the type of an answer that was required.

2.2.3 Data Requirements

2.2.3.1 Present or Near Term

Currently, data analysis, storage, and distribution efforts in EM are usually the responsibility of an individual staffer, or user; with whatever limited data analysis knowledge and capability available to them. As illustrated in Table 2.1, current

TABLE 2.1
Various Common Experiment Types and the Data Sizes for a Given Number of Probe Position (*x*, *y*) Coordinates and Higher Dimensional Energy Channels.

Experiment Type	Probe Positions	Pixels or Channels	Estimated Size (32-bit integer values)	Near Outlook (32-bit integer values)
Spectrum*	1	1 k	4 kB	
Ronchigram (Ronchi 1964)	1	1 × 1k	4 MB	16 MB
Line spectrum	128	1 k	512 kB	16 MB
Image**	1 × 1k	1	4 MB	64 MB (4 × 4k) × channels
Spectrum image	64 × 64	1 k	16 MB	16 GB (1 × 1 × 4k)
Ptychogram (Hoppe 1969)	64 × 64	256 × 256	1 GB	4 TB (1 × 1 × 1 × 1k)
Focal series	512 × 512	160	167 MB	1 GB (2k × 2k × 160)
Tilt series	1 × 1k	100	400 MB	1.6 GB (4k × 4k × 100)
Time series***	512 × 512	100	100 MB/frame	Many 4k × 4k × 100 frames (hours)

* Spectra could be electron energy loss spectra, x-ray spectra, or other detector feed.

** Current trends and short-term outlook for data generation and sizes in electron and ion beam microscopies.

*** These are intended to be "typical" values based on the ORNL systems that people are currently using, rather than the maximum possible. Depending on the sample, setup, or particular microscope hardware, values could easily increase by a factor of 2–4.

data volumes are already approaching the capacity for analysis on a local compute resource, such as a workstation computer. Soon small computational clusters will be necessary to handle even the simplest of operations in data visualization. In fact, this is the case for some stand-alone x-ray tomography instrumentation, as well as scanning EM (SEM). It is important to note here, that unlike physical probe microscopies [atomic force microscopy (AFM) and scanning tunneling microscopy (STM)] data generation time for EM and STEM are at least three orders of magnitude faster. On average, a single high-quality image on a physical probe microscope is collected in approximately 5 minutes at 512 × 512 pixels, whereas in a STEM/focused ion beam (FIB) a single 4 × 4k image can be captured well under a second. Images are perhaps the most basic, easy way to handle and process data types.

These data generation volumes extend beyond issues in processing and storage and in data transfer, particularly in experiments that rely on real-time feedback to the tool operator. This problem is complicated even further by the fact that many of the experiments summarized may happen concurrently with parallel data flows coming from independent detectors. Combined tilt-focal series, time series spectra, or through-focus ptychograms (Hoppe 1969), as well as movies that are even minutes in length, will take a great deal of space and require massive throughputs that necessitate livestreaming capabilities from the microscopes to efficiently transfer this information. It is immediately apparent from the near future trends in Figure 2.13 and Table 2.1 that these problems are only expected to get more severe in the near future.

Operating within a high-performance computing (HPC), or cloud environment, will provide the key interface for intimate interaction of experiment and theory. The multimodal, hyperspectral data collected with new generation microscopy techniques are an amalgam of what is typically independently processed by well-established, theoretical techniques that utilize self-contained approaches, but they are rarely cross-validated. These independent analysis workflows are well understood and widely utilized in a high-intensive computational environment by theoreticians today. Combining storage, preprocessing, and theoretical efforts will intertwine experiment and theory into a single, streamlined analysis process enabled by an HPC environment. Naturally, the grand goal of uniting these efforts is to enable true theoretical feedback to guide experiment and discovery in near real time.

2.2.3.2 Data Lifecycle

The data lifecycle follows a familiar cyclical pattern commonly found in data life management literature and replicated in Figure 2.14.

- *Creating Data:* The microscope generates almost all of the data. There is some additional descriptor metadata associated with the sample, operator, and microscope state, but it is usually infinitesimal compared with the size of the detector output. The number of detectors can vary but will rarely cross over into double digits. The raw data output is uncompressed and is currently typically at 32-bit integers (software limited), with older microscope detectors clamped at 16 bits. The data transfer mechanism from the detector to the storage media depends on the manufacturer, but the most commonly used interfaces are USB, Ethernet, and PCI/PCI-e.

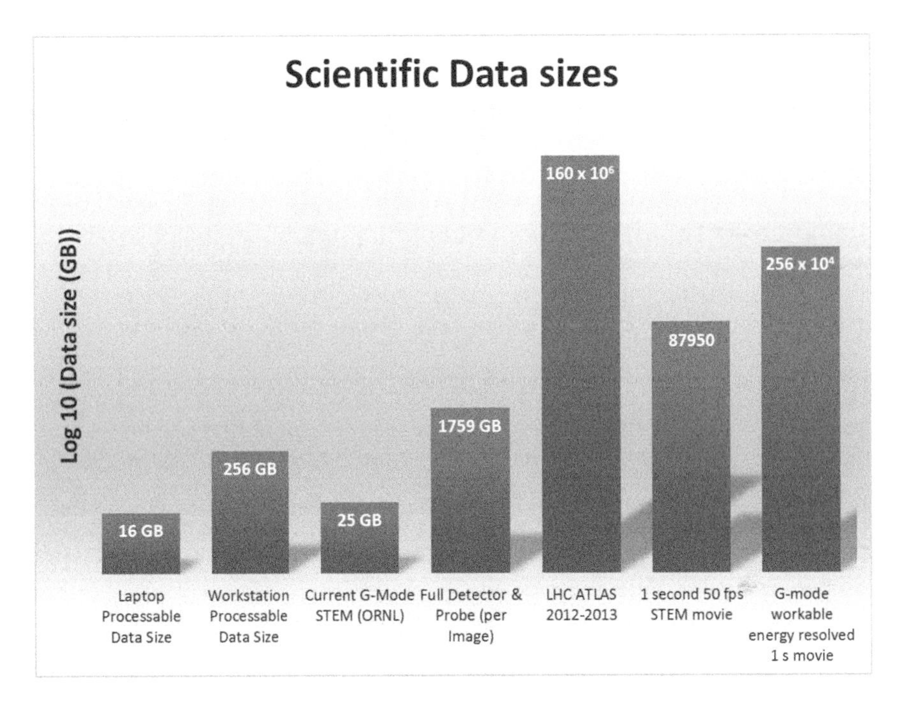

FIGURE 2.13 Scientific data sizes on the processing and generation ends are depicted. Laptop and workstation capabilities are estimated by average machines available on the market today, such as the 25 GB G-mode (Belianinov et al. 2015). STEM is a single, small (Table 2.1) 4D (200 × 200 × 400 × 400) dataset. Full detector and probe data size is for a single 4D hyperspectral dataset where the output of all electron or ion probe positions (2048 × 2048) is captured at an *average size* detector array (2048 × 2048). Large hardon collider (LHC) ATLAS detector output is shown for 2012–2013. A 50-frame-per-second movie captured at 2048 × 2048 probe positions with a 2048 × 2048 detector array data size is seen in this figure. The data size for a modest G-mode movie captured at 200 × 200 probe positions on a 768 × 768-pixel detector (per one exposure), with 256 Ronchigram energy channels over 40 frames, is also illustrated.

- *Processing Data:* Classical processing methods in STEM and FIB utilize binning and averaging (or lock-in amplification) to improve the signal to noise; however, over the last few decades detection hardware has improved significantly, with the processing methods largely utilized to control data volumes. At the very first stages of the analysis workflow we are interested in collecting full detector response at fastest meaningful rates to assess tool performance and adjust parameters on the fly. There is some backlash to this mode of thinking, as there have been demonstrations that the same material insight can be generated with downsampled data. (Jesse et al. 2016) Unfortunately, this argument can only be made a posteriori, and it seems unwise to throttle full hardware capability of usually very expensive new detector hardware to save a little on storage. Additionally, fast visualization schemes would be of use to monitor the sample and the quality of the output signal.

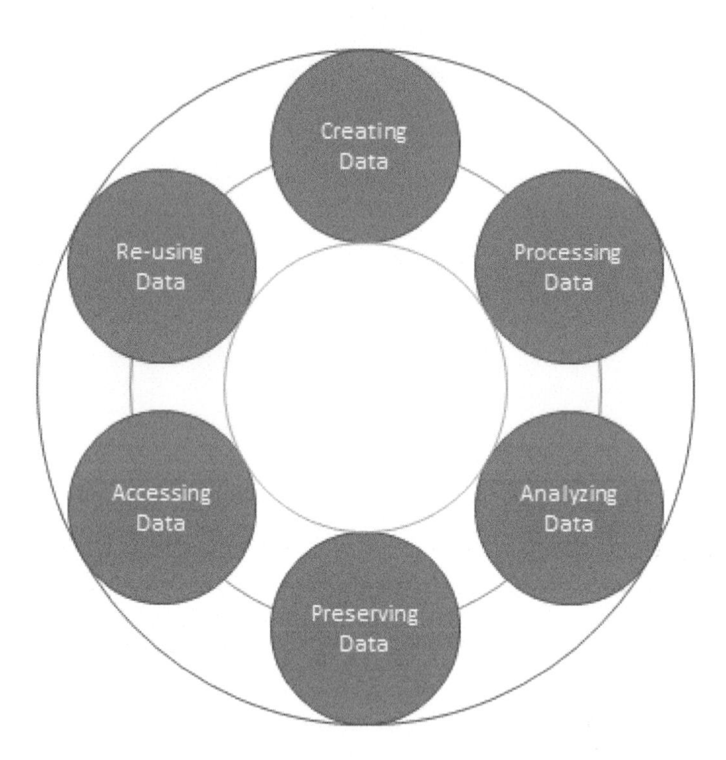

FIGURE 2.14 Data lifecycle process reusing EM.

- *Analyzing Data:* The analysis framework must be scalable, parallelizable, and flexible. Due to the maturity of the field, a large number of instrumental configurations, and breadth of scientific interests, the analytical back end has to have the capability to adapt to either completely new analysis library software, or have the flexibility to combine analysis workflows in an unconstrained fashion. Currently the analysis requirements include standard packages for plotting, 2/3D visualization, File IO, and mathematical and multivariate statistical processing libraries. Soon more exotic DNN, image and hyperspectral registration and segmentation, compressed sensing, and robust file sharing packages will be necessary.
- *Preserving Data:* Data stewardship must be included with the data collection and processing services. Data quotas, length of storage, redundancy, and encryption are only some of the key details that must be discussed, in the earliest project proposal stages. This is the missing puzzle piece to address the missing standardized data criticism often aimed at complex learning methods. However, some form of basic, short-term storage and access over the lifetime of the user project for pertinent analyzed and raw datasets must be available.
- *Accessing Data:* In the current framework the data belong to the supporting agency. Associated students, staff, and co-PIs have access to the data

with the permission of the main PI. Data access during its lifetime on the compute resources should be limited to the individuals with proper training and security credentials. This is complicated by the framework of open data access creating storage of redundant information without clear guidelines on who should be supporting the infrastructure.

- *Reusing Data:* Due to the nature of the experiments and the statistical framework to analyze and refine the data, recurring analysis of the same dataset is vital for understanding the underlying physics and chemistry, as well as validity of proposed theoretical models. We expect that scientists will reanalyze their data multiple times, and in time will utilize the data file structures that capture and contain the results of periodic reanalysis information. Additionally, the results of such a persistent approach to analysis can be cross-correlated and form some of the very bases of scientific data standards that are currently lacking in many areas.

2.2.3.3 Data-Centric Requirements: Capabilities, Speeds, and Feeds

Table 2.1 summarized various common experiment types and the data sizes for a given number of probe position (x, y) coordinates and higher dimensional energy channels. Looking beyond proposed values to a 5- to 10-year outlook, the data sizes will continue to grow. Realistically, for each electron, we could record the probe position (x, y), scattering angle (u, v), and the energy loss (E, here we assume that you record the energy loss as a single value, not as a complete spectrum); resulting in a 5D dataset as the most basic data unit. Additionally, these data could be a function of frame or focus or some physical parameter (ptychogram, focal series, tilt series, etc.) adding dimensionality and size. Recording (x, y, u, v, E) gives roughly 20 bytes of data (using 32-bit integer values) per electron, providing a data rate of 4 GB/s, for a standard imaging case of approximately 32-pA current, which is roughly 200 electrons per microsecond per detector.

2.2.3.4 Impediments, Gaps, Needs, Challenges

We face the following three serious impediments blocking the success of developing the field EM in the information age:

1. Real-time processing and high data throughput to support real-time/near real-time experimental feedback, with data transfer rates of at least 4 GB/s required to address this challenge.
2. Large, accessible data repository for archival and data sharing, with an annual storage capacity of 5 PB.
3. Limited access to a sufficient computing resource for the analysis of generated data; with a dedicated ~2000 (64 per microscopy tool) node system with a graphics processing unit (GPU) capability required to address this challenge.

While each of the aforementioned roadblocks is a challenging issue at even the current rates of data generation and analysis, what hinders the future of EM can be summarized as the *lack of resources to move, store, share, and process scientific data*. According to the even modest near future estimates outlined in Table 2.1, continuous, secure data transfer rates of approximately 4 GB/s *per microscope* are required to sustain tool operation and adequately fulfill a center's obligation to the user community. Furthermore, global data accessibility is an important preamble to analysis, as input from various experts in the field is vital to achieve real scientific progress. Therefore, a flexible data repository that allows fast and secure access to data by a handful of individuals to advise and oversee the analysis process is critical. Finally, the ability to receive real-time, or near real-time feedback to the operator requires a dedicated computational resource to each microscope. Soon, satisfying both data-processing and theory requirements for the data stream off each microscope is likely unrealistic; however, previewing the analyzed data and getting it ready for serious theoretical effort is an excellent near future goal.

The emerging trends place heavy emphasis on combinatorial imaging that correlates spatial, chemical, and physical information. Serious challenges in processing these data have been slowly, but steadily addressed by the scientific community on many fronts; however, scaling, validating, and cross-correlating these independent efforts is a serious roadblock that can only be addressed by close collaboration with the HPC and data handling experts. We foresee that close ties with the information technology community would usher in a technical revolution in the scientific fronts by providing the infrastructure for truly close-knit multidiscipline collaboration.

2.3 OUTLOOK

There are without a doubt many opportunities for ML in EM. The nearest term substantial changes include automation and a closer integration with theory, or at least much improved accuracy. The most interesting, on the hardware-software interface front, is to place the microscope into a feedback loop between the machine and the deep learning framework trained for specific imaging applications. This would help to set tool parameters and can aid in automated characterization of materials. This triggers a redesign in the experimental approaches in which the microscopes start acting as semiautonomous agents, capable of processing similar sample tasks all the way from loading the specimens to automatically generating a report on parameters of interest.

A critical concern in utilizing deep learning for science is accuracy. This is an open question as there is a dearth of open standardized datasets for EM in which algorithms can be tested and evaluated for bias, error, and robustness. One of the most promising approaches to improve accuracy right now is to start utilizing a combination of deep learning and Gaussian processes, such as neural processes (Garnelo et al. 2018a,b), in addition Bayesian inference can also be used. (Mueller, Kusne and Ramprasad 2016; Dyck et al. 2018a). Undoubtedly, as the methodology matures and becomes more widespread algorithm standards and/or standard datasets will begin to percolate.

Whatever the future holds, EM will surely benefit from the technological developments of the information age. These changes may not come easy or quick, as ideas take time to root and resistance to implementation needs to recede. Nonetheless, curiosity and excitement associated with visualizing some of the smallest parts of our universe will undoubtedly force progress. Just think about that feeling the first time you collected an EM image and in a flash realized, *these are atoms!*

ACKNOWLEDGMENTS

This work was supported by the Center for Nanophase Materials Sciences, which is a Department of Energy Office of Science User Facility.

REFERENCES

Allen, G., and Chan, T., 2017. *Artificial Intelligence and National Security*, Belfer Center for Science and International Affairs.

Angermueller, C., Pärnamaa, T., Parts, L., and Stegle, O., 2016. Deep learning for computational biology. *Molecular Systems Biology*, *12*(7): 878.

Arthur, D., and Vassilvitskii, S., k-means++: The advantages of careful seeding. Proceedings of the 18th annual ACM-SIAM symposium on Discrete algorithms, New Orleans, Louisiana, 2007.

Baldi, P., Bauer, K., Eng, C., Sadowski, P., and Whiteson, D., 2016. Jet substructure classification in high-energy physics with deep neural networks. *Physical Review D*, *93*(9): 094034.

Baldi, P., Sadowski, P., and Whiteson, D., 2014. Searching for exotic particles in high-energy physics with deep learning. *Nature Communications*, *5*: 4308.

Belianinov, A., He, Q., Kravchenko, M., Jesse, S., Borisevich, A., and Kalinin, S.V., 2015. Identification of phases, symmetries and defects through local crystallography. *Nature Communications*, *6*: 7801.

Belianinov, A., Kalinin, S.V., and Jesse, S., 2015. Complete information acquisition in dynamic force microscopy. *Nature Communications*, *6*(1): 1–7.

Belianinov, A., Vasudevan, R., Strelcov, E., Steed, C., Yang, S.M., Tselev, A., Jesse, S., Biegalski, M., Shipman, G., Symons, C., and Borisevich, A., 2015. Big data and deep data in scanning and electron microscopies: deriving functionality from multidimensional data sets. *Advanced Structural and Chemical Imaging*, *1*(1): 6.

Borisevich, A., Ovchinnikov, O.S., Chang, H.J., Oxley, M.P., Yu, P., Seidel, J., Eliseev, E.A., Morozovska, A.N., Ramesh, R., Pennycook, S.J., and Kalinin, S.V., 2010a. Mapping octahedral tilts and polarization across a domain wall in $BiFeO_3$ from Z-contrast scanning transmission electron microscopy image atomic column shape analysis. *ACS Nano*, *4*(10): 6071–6079.

Borisevich, A.Y., Chang, H.J., Huijben, M., Oxley, M.P., Okamoto, S., Niranjan, M.K., Burton, J.D., Tsymbal, E.Y., Chu, Y.H., Yu, P. and Ramesh, R., 2010v. Suppression of octahedral tilts and associated changes in electronic properties at epitaxial oxide heterostructure interfaces. *Physical Review Letters*, *105*(8): 087204.

Bruner, J.S., Goodnow, J.J., and Austin, G.A., 1956. *A Study of Thinking*, Science Editions. Inc.

Campello, R.J., Moulavi, D., Zimek, A. and Sander, J., 2015. Hierarchical density estimates for data clustering, visualization, and outlier detection. *ACM Transactions on Knowledge Discovery from Data (TKDD)*, *10*(1): 1–51.

Chang, H.J., Kalinin, S.V., Morozovska, A.N., Huijben, M., Chu, Y.H., Yu, P., Ramesh, R., Eliseev, E.A., Svechnikov, G.S., Pennycook, S.J., and Borisevich, A.Y., 2011. Atomically resolved mapping of polarization and electric fields across Ferroelectric/Oxide interfaces by z-contrast imaging. *Advanced Materials*, *23*(21): 2474–2479.

Dan, J., Zhao, X., and Pennycook, S.J., 2019. A machine perspective of atomic defects in scanning transmission electron microscopy. *InfoMat.*, *1*(3): 359–375.

Dasgupta, S., and Freund, Y., Random projection trees and low dimensional manifolds. Proceedings of the 40th annual ACM symposium on Theory of computing. Victoria, British Columbia, 2008.

Dong, W., Moses, C., and Li, K., 2011. Efficient k-nearest neighbor graph construction for generic similarity measures. In *Proceedings of the 20th International Conference on World Wide Web* (pp. 577–586).

Drucker, P., 1994. The theory of the business. Harvard Business Review, October.

Dyck, O., Bao, F., Ziatdinov, M., Nobakht, A.Y., Shin, S., Law, K., Maksov, A., Sumpter, B.G., Archibald, R., Jesse, S., and Kalinin, S.V., 2018a. Single atom force measurements: mapping potential energy landscapes via electron beam induced single atom dynamics. *arXiv preprint arXiv:1804. 03729.*

Dyck, O., Kim, S., Kalinin, S.V., and Jesse, S., 2018b. E-beam manipulation of Si atoms on graphene edges with an aberration-corrected scanning transmission electron microscope. *Nano Research*, *11*(12): 6217–6226.

Garnelo, M., Rosenbaum, D., Maddison, C.J., Ramalho, T., Saxton, D., Shanahan, M., Teh, Y.W., Rezende, D.J., and Eslami, S.M., 2018a. Conditional neural processes. *arXiv preprint arXiv:1807. 01613.*

Garnelo, M., Schwarz, J., Rosenbaum, D., Viola, F., Rezende, D.J., Eslami, S.M., and Teh, Y.W., 2018b. Neural processes. *arXiv preprint arXiv:1807. 01622.*

Glorot, X., Bordes, A., and Bengio, Y., Deep sparse rectifier neural networks. Proceedings of the 14th International Conference on Artificial Intelligence and Statistics. 2011, Ft. Landerdale, FL, USA.

Gómez-Bombarelli, R., Wei, J.N., Duvenaud, D., Hernández-Lobato, J.M., Sánchez-Lengeling, B., Sheberla, D., Aguilera-Iparraguirre, J., Hirzel, T.D., Adams, R.P., and Aspuru-Guzik, A., 2018. Automatic chemical design using a data-driven continuous representation of molecules. *ACS Central Science*, *4*(2): 268–276.

Gong, M.L., Miller, B.D., Unocic, R.R., Hattar, K., Reed, B., Masiel, D., Tasdizen, T. and Aguiar, J.A., 2019. Merging deep learning, chemistry, and diffraction for high-throughput material structure prediction. *Microscopy and Microanalysis*, *25*(S2): 168–169.

Goodfellow, I., Bengio, Y., and Courville, A., 2016. *Deep Learning*. Vol. 1: MIT Press.

Green, M.L., Choi, C.L., Hattrick-Simpers, J.R., Joshi, A.M., Takeuchi, I., Barron, S.C., Campo, E., Chiang, T., Empedocles, S., Gregoire, J.M. and Kusne, A.G., 2017. Fulfilling the promise of the materials genome initiative with high-throughput experimental methodologies. *Applied Physics Reviews*, *4*(1): 011105.

Guest, D., Collado, J., Baldi, P., Hsu, S.C., Urban, G. and Whiteson, D., 2016. Jet flavor classification in high-energy physics with deep neural networks. *Physical Review D*, *94*(11): 112002.

Hattrick-Simpers, J.R., Gregoire, J.M. and Kusne, A.G., 2016. Perspective: composition–structure–property mapping in high-throughput experiments: turning data into knowledge. *APL Materials*, *4*(5): 053211.

He, J., Borisevich, A., Kalinin, S.V., Pennycook, S.J., and Pantelides, S.T., 2010. Control of octahedral tilts and magnetic properties of perovskite oxide heterostructures by substrate symmetry. *Physical Review Letters*, *105*(22): 227203.

He, Q., Woo, J., Belianinov, A., Guliants, V.V. and Borisevich, A.Y., 2015. Better catalysts through microscopy: mesoscale M1/M2 intergrowth in Molybdenum–Vanadium based complex oxide catalysts for propane ammoxidation. *ACS Nano*, *9*(4): 3470–3478.

Holdren, J.P., 2011. Materials Genome Initiative for Global Competitiveness. *National Science and Technology Council OSTP. Washington, USA.*

Hoppe, W., 1969. Beugung im inhomogenen primärstrahlwellenfeld. I. Prinzip einer phasenmessung von elektronenbeungungsinterferenzen. *Acta Crystallographica Section A: Crystal Physics, Diffraction, Theoretical and General Crystallography*, *25*(4): 495–501.

Hyvärinen, A. and Oja, E., 2000. Independent component analysis: algorithms and applications. *Neural Networks*, *13*(4–5): 411–430.

Jain, A., Ong, S.P., Hautier, G., Chen, W., Richards, W.D., Dacek, S., Cholia, S., Gunter, D., Skinner, D., Ceder, G. and Persson, K.A., 2013. Commentary: the materials project: a materials genome approach to accelerating materials innovation. *APL Materials*, *1*(1): 011002.

Jesse, S., Chi, M., Belianinov, A., Beekman, C., Kalinin, S.V., Borisevich, A.Y. and Lupini, A.R., 2016. Big data analytics for scanning transmission electron microscopy ptychography. *Scientific Reports* *6*:26348. https://www. nature. com/articles/srep26348#supplementary-information.

Jia, C.L., Mi, S.B., Faley, M., Poppe, U., Schubert, J., and Urban, K., 2009. Oxygen octahedron reconstruction in the SrTiO(3)/LaAlO(3) heterointerfaces investigated using aberration-corrected ultrahigh-resolution transmission electron microscopy. *Physical Review B*, *79*: 081405(R).

Jia, C.L., Nagarajan, V., He, J.Q., Houben, L., Zhao, T., Ramesh, R., Urban, K. and Waser, R., 2007. Unit-cell scale mapping of ferroelectricity and tetragonality in epitaxial ultrathin ferroelectric films. *Nature Materials*, *6*(1): 64–69.

Jia, C.L., Urban, K.W., Alexe, M., Hesse, D. and Vrejoiu, I., 2011. Direct observation of continuous electric dipole rotation in flux-closure domains in ferroelectric Pb(Zr,Ti)O(3). *Science*, *331*(6023): 1420–1423.

Jones, L., MacArthur, K.E., Fauske, V.T., van Helvoort, A.T. and Nellist, P.D., 2014. Rapid estimation of catalyst nanoparticle morphology and atomic-coordination by high-resolution Z-contrast electron microscopy. *Nano Letters*, *14*(11): 6336–6341.

Jones, L. and Nellist, P.D., 2013. Identifying and correcting scan noise and drift in the scanning transmission electron microscope. *Microscopy and Microanalysis*, *19*(4): 1050–1060.

Kalinin, S.V., Lupini, A.R., Dyck, O., Jesse, S., Ziatdinov, M. and Vasudevan, R.K., 2019. Lab on a beam—Big data and artificial intelligence in scanning transmission electron microscopy. *MRS Bulletin*, *44*(7): 565–575.

Kannan, R., Ievlev, A.V., Laanait, N., Ziatdinov, M.A., Vasudevan, R.K., Jesse, S. and Kalinin, S.V., 2018. Deep data analysis via physically constrained linear unmixing: universal framework, domain examples, and a community-wide platform. *Advanced Structural and Chemical Imaging*, *4*(1): 6.

Kim, Y.M., He, J., Biegalski, M.D., Ambaye, H., Lauter, V., Christen, H.M., Pantelides, S.T., Pennycook, S.J., Kalinin, S.V. and Borisevich, A.Y., 2012. Probing oxygen vacancy concentration and homogeneity in solid-oxide fuel-cell cathode materials on the subunit-cell level. *Nature Materials*, *11*(10): 888–894.

Kim, Y.M., Kumar, A., Hatt, A., Morozovska, A.N., Tselev, A., Biegalski, M.D., Ivanov, I., Eliseev, E.A., Pennycook, S.J., Rondinelli, J.M. and Kalinin, S.V., 2013. Interplay of octahedral tilts and polar order in $BiFeO_3$ films. *Advanced Materials*, *25*(17): 2497–2504.

Kim, Y.M., Morozovska, A., Eliseev, E., Oxley, M.P., Mishra, R., Selbach, S.M., Grande, T., Pantelides, S.T., Kalinin, S.V. and Borisevich, A.Y., 2014. Direct observation of ferroelectric field effect and vacancy-controlled screening at the $BiFeO_3$/LaxSr1-xMnO_3 interface. *Nature Materials*, *13*(11): 1019–1025.

Laanait, N., Ziatdinov, M., He, Q. and Borisevich, A., 2016. Identifying local structural states in atomic imaging by computer vision. *Advanced Structural and Chemical Imaging*, *2*(1): 14.

LeBeau, J.M., Findlay, S.D., Allen, L.J. and Stemmer, S., 2008. Quantitative atomic resolution scanning transmission electron microscopy. *Physical Review Letters*, *100*(20): 206101.

LeBeau, J.M., Findlay, S.D., Allen, L.J. and Stemmer, S., 2010. Standardless atom counting in scanning transmission electron microscopy. *Nano Letters*, *10*(11): 4405–4408.

Lee, D.D. and Seung, H.S., 1999. Learning the parts of objects by non-negative matrix factorization. *Nature*, *401*(6755): 788.

Li, Q., Nelson, C.T., Hsu, S.L., Damodaran, A.R., Li, L.L., Yadav, A.K., McCarter, M., Martin, L.W., Ramesh, R. and Kalinin, S.V., 2017. Quantification of flexoelectricity in $PbTiO_3/SrTiO_3$ superlattice polar vortices using machine learning and phase-field modeling. *Nature Communications*, *8*(1): 1468.

Li, X., Dyck, O.E., Oxley, M.P., Lupini, A.R., McInnes, L., Healy, J., Jesse, S. and Kalinin, S.V., 2019. Manifold learning of four-dimensional scanning transmission electron microscopy. *NPJ Computational Materials*, *5*(1): 1–8.

Litjens, G., Kooi, T., Bejnordi, B.E., Setio, A.A.A., Ciompi, F., Ghafoorian, M., Van Der Laak, J.A., Van Ginneken, B. and Sánchez, C.I., 2017. A survey on deep learning in medical image analysis. *Medical Image Analysis*, *42*: 60–88.

Lu, P. and Gauntt, B.D., 2013. Structural mapping of disordered materials by nanobeam diffraction imaging and multivariate statistical analysis. *Microscopy and Microanalysis*, *19*(2): 300–309.

Maksov, A., Dyck, O., Wang, K., Xiao, K., Geohegan, D.B., Sumpter, B.G., Vasudevan, R.K., Jesse, S., Kalinin, S.V. and Ziatdinov, M., 2019. Deep learning analysis of defect and phase evolution during electron beam-induced transformations in WS2. *NPJ Computational Materials*, *5*(1): 12.

Marcus, G., Rossi, F. and Veloso, M., 2016. Beyond the turing test. *AI Magazine*, *37*(1): 3–4.

McInnes, L., Healy, J. and Astels, S., 2017. Hdbscan: hierarchical density based clustering. *Journal of Open Source Software*, *2*(11): 205.

McInnes, L., Healy, J., and Melville, J., 2018. Umap: uniform manifold approximation and projection for dimension reduction. *arXiv preprint arXiv:1802. 03426.*

Mikolov, T., Sutskever, I., Chen, K., Corrado, G.S., and Dean, J., 2013. Distributed representations of words and phrases and their compositionality. *arXiv preprint arXiv:1310. 04546.*

Mody, C., 2011. *Instrumental Community: Probe Microscopy and the Path to Nanotechnology*, MIT Press.

Moore, G.E., 1998. Cramming more components onto integrated circuits. *Proceedings of the IEEE*, *86*(1): 82–85.

Mueller, T., Kusne, A.G. and Ramprasad, R., 2016. Machine learning in materials science: recent progress and emerging applications. *Reviews in Computational Chemistry*, 29: 186–273.

Nair, V., and Hinton, G.E., Rectified linear units improve restricted boltzmann machines. Proceedings of the 27th International Conference on Machine Learning (ICML-10) 2010, Haifa, Israel.

Najem, J.S., Taylor, G.J., Weiss, R.J., Hasan, M.S., Rose, G., Schuman, C.D., Belianinov, A., Collier, C.P. and Sarles, S.A., 2018. Memristive ion channel-doped biomembranes as synaptic mimics. *ACS Nano*, *12*(5): 4702–4711.

Nelson, C.T., Winchester, B., Zhang, Y., Kim, S.J., Melville, A., Adamo, C., Folkman, C.M., Baek, S.H., Eom, C.B., Schlom, D.G. and Chen, L.Q., 2011. Spontaneous vortex nanodomain arrays at ferroelectric heterointerfaces. *Nano Letters*, *11*(2): 828–834.

Oh, K.S. and Jung, K., 2004. GPU implementation of neural networks. *Pattern Recognition*, *37*(6): 1311–1314.

Oliner, S.D., and Sichel, D.E., 2002. *Information Technology and Productivity: Where Are We Now and Where Are We Going?*, Divisions of Research & Statistics and Monetary Affairs, Federal Reserve Board.

Pennycook, S.J., Chisholm, M.F., Lupini, A.R., Varela, M., Van Benthem, K., Borisevich, A.Y., Oxley, M.P., Luo, W., and Pantelides, S.T., 2008. Materials applications of aberration-corrected scanning transmission electron microscopy. In *Advances in Imaging and Electron Physics, Vol 153*, edited by P. W. Hawkes, 327-+. Elsevier Academic Press Inc.

Pennycook, S.J., and Nellist, P.D., eds. 2011. *Scanning Transmission Electron Microscopy: Imaging and Analysis*, Springer.

Ronchi, V., 1964. Forty years of history of a grating interferometer. *Applied Optics*, *3*(4): 437–451.

Roweis, S.T., and Saul, L.K., 2000. Nonlinear dimensionality reduction by locally linear embedding. *Science*, *290*(5500): 2323–2326.

Russell, S.J., and Norvig, P., 2016. *Artificial Intelligence: a Modern Approach*, Pearson Education Limited.

Sarahan, M.C., Chi, M., Masiel, D.J. and Browning, N.D., 2011. Point defect characterization in HAADF-STEM images using multivariate statistical analysis. *Ultramicroscopy*, *111*(3): 251–257.

Schubert, E., Sander, J., Ester, M., Kriegel, H.P. and Xu, X., 2017. DBSCAN revisited, revisited: why and how you should (still) use DBSCAN. *ACM Transactions on Database Systems (TODS)*, *42*(3): 19.

Singh, A.P., and Gordon, G.J., 2008. A unified view of matrix factorization models. In *Joint European Conference on Machine Learning and Knowledge Discovery in Databases* (pp. 358–373). Springer.

Strelcov, E., Belianinov, A., Hsieh, Y.H., Jesse, S., Baddorf, A.P., Chu, Y.H. and Kalinin, S.V., 2014. Deep data analysis of conductive phenomena on complex oxide interfaces: physics from data mining. *ACS Nano*, *8*(6): 6449–6457.

Susi, T., Meyer, J.C. and Kotakoski, J., 2017. Manipulating low-dimensional materials down to the level of single atoms with electron irradiation. *Ultramicroscopy*, *180*: 163–172.

Tang, J., Liu, J., Zhang, M., and Mei, Q., Visualizing large-scale and high-dimensional data. Proceedings of the 25th International Conference on World Wide Web 2016 Montréal Québec Canada.

Treverton, G., 2017. Global trends: paradox of progress. *NIC/DNI, Washington*, *vi–ix*.

Vasudevan, R.K., Belianinov, A., Gianfrancesco, A.G., Baddorf, A.P., Tselev, A., Kalinin, S.V. and Jesse, S., 2015. Big data in reciprocal space: sliding fast fourier transforms for determining periodicity. *Applied Physics Letters*, *106*(9): 091601.

Yankovich, A.B., Berkels, B., Dahmen, W., Binev, P., Sanchez, S.I., Bradley, S.A., Li, A., Szlufarska, I. and Voyles, P.M., 2014. Picometre-precision analysis of scanning transmission electron microscopy images of platinum nanocatalysts. *Nature Communications*, *5*(1): 1–7.

Ziatdinov, M., Dyck, O., Maksov, A., Li, X., Sang, X., Xiao, K., Unocic, R.R., Vasudevan, R., Jesse, S. and Kalinin, S.V., 2017. Deep learning of atomically resolved scanning transmission electron microscopy images: chemical identification and tracking local transformations. *ACS Nano*, *11*(12): 12742–12752.

Zurada, J.M., 1992. *Introduction to Artificial Neural Systems*, Vol. 8, West Publishing Company.

3 Application of Advanced Aberration-Corrected Transmission Electron Microscopy to Material Science: Methods to Predict New Structures and Their Properties

O. I. Lebedev
Laboratoire CRISMAT, UMR 6508 CNRS/ENSICAEN/UCBN,
6 bd du Maréchal Juin, F-14050 CAEN Cedex 4 – France

CONTENTS

3.1 Introduction 70
3.2 Method to Characterize Materials by Means of Electron Microscopy: Case Studies 75
3.2.1 The Manganoferrite $Pb_{2-x}Ba_xFeMnO_5$ 75
3.2.2 Rare-earth Cations Inside BaLaCuP Clathrate Polyhedral Cages Revealed by Advanced TEM 78
3.2.3 CuInSe Hexagonal Flat Nanoparticles 80
3.2.4 Revisiting Hollandite $Bi_{2-x}V_yV_8O_{16}$ Cubic Structure Thanks to HAADF-STEM 84
3.3 Advanced Tem Characterization of Perovskite Materials: A Straightforward Method to Analyze Novel Cation Ordered Materials 90
3.3.1 Quintuple Perovskites Revealed by Advanced TEM 91
3.3.1.1 Cation Ordering in $Sm_{2-\varepsilon}Ba_{3+\varepsilon}$ Fe_5 $O_{15-\delta}$ Complex Perovskite 91
3.3.1.2 Complex Ordering of Eu and Ba in $Eu_2Ba_3Fe_3Co_2O_{15-\delta}$ Quintuple Oxygen-Deficient Perovskite 95
3.3.2 Exceptional Layered Ordering of Cobalt and Iron in $Y_2Ba_3Fe_3Co_2O_{13.36}$ Oxygen-Deficient Perovskite Revealed by Advanced TEM 101
References 105

3.1 INTRODUCTION

The history of transmission electron microscopy (TEM) starts from the 1930s when Ernst Ruska invented and realized the first transmission electron microscope (Ruska 1933). Since then, the development of electron microscopy techniques and microscopes goes hand in hand with the discovery of new materials. Among the numerous powerful and widely used techniques for solid materials characterization, such as x-ray diffraction (XRD), Mossbauer and Raman spectroscopy, and so forth, TEM was to a certain extent a supplementary characterization technique until the 1970s. However, with the development of the electron microscope as an instrument, increasing stability and resolution, TEM becomes more and more significant and valuable for materials science. In the last 15 years, the progress in transmission electron microscope techniques made TEM almost compulsory in the studies of novel materials. Each of these TEM techniques can be applied to a number of problems, but very often, even for the specialists, it is not evident which techniques should be used among all numerous methods to answer particular materials science questions. A detailed discussion on each technique is not the aim of this chapter; however, it is quite useful to summarize the most known techniques and their applications toward understanding and answering various problems regarding materials structure characterization; Table 3.1 aims to do this. Along with the particular indication of which information (structural, chemical, electronic, etc.) can be obtained with different techniques, some references are given in Table 3.1, where more practical and theoretical information can be found.

Conventional TEM (CTEM) uses electrons focused by electromagnetic lenses into a fine electron beam. Due to the high accelerating voltage use in the electron microscope, the short wavelength of the accelerated electrons (around 2 pm for a 300-kV accelerating voltage) made the resolution of an electron microscope significantly high to solve structural problems in various materials. The classical set of TEM techniques for primary material characterization was electron diffraction (ED), bright-field (BF) imaging and high-resolution TEM (HRTEM). At the same time, energy-dispersive x-ray spectroscopy (EDXS) offers local chemical information. ED patterns collected from main crystallographic zone axes give basic information about the crystal structure. The angles between diffraction spots, showing the presence or absence of certain reflections, can be used to determine the crystal structure (unit cell parameters, extinction conditions, space groups) and can be correlated with data obtained from XRD (Fultz and Howe 2013).

The shape of the diffraction spots, the presence of streaks, and weak superstructure spots are characteristic of imperfections in materials and might be arising due to different kinds of defects, twinning, and superstructure ordering, which are not detectable by XRD techniques. In spite of the different natures of ED and XRD (the nature of electrons is dynamical, while that of x-rays is kinematical), the two techniques are coupled; ED becomes like a bridge between the nanoscale (in the case of TEM) and macroscale structures (in the case of XRD). Moreover, information on possible weak superstructures (commensurate or noncommensurate), twinning, extended defects or local distortions, and selective area ED (SAED) from nanoparticles (NPs) can be detected only by ED. Therefore, for certain complex nanostructures, it becomes necessary to extract crystallographic information.

Diffraction and imaging in an electron microscope go "hand-in-hand": all features of ED patterns will be present in TEM images, and they are related by the Fourier transform (FT) from reciprocal diffraction space to real space in image mode. In other words, the image and the diffraction pattern of a certain crystal structure are each other's FT. The standard and basis operation mode in CTEM is the BF imaging and consists of mass-thickness and diffraction contrast imaging (Williams and Carter 1996). Mass-thickness contrast, also known as scattering or absorption contrast, arises from incoherently (Rutherford) elastically scattered electrons and is dominant for noncrystalline materials. In crystalline materials, this amplitude-type contrast is in direct competition with the diffraction contrast, which originates from elastic Bragg scattering. In this case, the image and contrast of the structure along a specific zone axis depend on the exact orientation of the crystalline specimen with respect to the incoming electron beam. This allows the imaging of the crystal structure at an atomic scale.

An HRTEM image is obtained by the interference of the direct incident transmitted electron beam and diffracted beams. As a result, the contrast in such an image is dependent on phase of the various beams and is known as phase-contrast imaging. The electron scattering is strongly dynamical; therefore, multiple scattering leads to large phase changes. The exit wave, the electron wave at the exit plane of the crystal, has all structural information transferring through all lenses of the microscope. The transfer function in the electron microscope is not ideal due to different lens aberrations (spherical, chromatic, astigmatism, coma, defocus, etc.) and frequencies present in the exit wave will act differently. Hence, direct interpretation of the atomic arrangement in HRTEM images is seldom possible. In this respect, a proper interpretation of HRTEM images is only aided by using image simulation, multislice or Bloch wave approximations that take parameters that can affect the HRTEM image, such as defocus, lens aberration, thickness, multiple scattering, and so forth, into account. Therefore, the combination of ED, HRTEM, and EDXS techniques is a powerful and reliable set to solve the structure of novel inorganic materials, nano materials, and thin films (Van Tendeloo et al. 2004; Kirkland 2010; Fultz and Howe 2013).

However, until the end of the last century, the limited factors for CTEM image resolution were the spherical and chromatic aberrations caused by electromagnetic lenses (Scherzer 1936). The best point resolution at optimal Scherzer defocus was around 1.7 Å at 400 kV, which is far from the theoretical resolution of only a few picometers. In the 1990s, spherical aberration correction became a reality in TEM, (Haider et al. 1998; Krivanek, Dellby and Lupini 1999) causing a revolution in the field and pushing development of new and even already known techniques to the state-of-art instrumentation and new TEM techniques [e.g. atomic manipulation and lattice engineering single atom dynamics using electron irradiation (Su et al. 2019; Dyck, Kim and Kalinin 2017), or vortex electron beams TEM (Verbeeck, Tian and Schattschneider 2010)]. It was found that spherical aberration can be corrected using either a hexapole (Beck 1979; Rose 1990) or a quadrupole-octupole (Rose 1971a,b) aberration corrector lens to create a negative spherical aberration, which will effectively cancel out the original positive spherical aberration.

TABLE 3.1
Overview of TEM Techniques for Different Materials Science Needs

Information on	TEM Techniques	References
Crystal Structure	Electron diffraction (ED)	Williams and Carter 1996; Cowley 1992a; Amelinckx et al. 1997; DeGraef 2002
	Convergent beam ED (CBED)	Scherzer 1936; Kossel and Möllenstedt 1939; Goodman and Lehmpfuhl 1968; Champnes 2001
	Precession electron diffraction (PED) High-resolution TEM (HRTEM)	Morniroli and Steeds 1992; Vincent and Midgley 1994
	HAADF-STEM	Scherzer 1936; Shindo and Spence 1981; Hiraga 1998; Van Tendeloo et al. 2004
	ABF-STEM	Pennycook 1992; Findlay et al. 2009; Erni 2010; Pennycook and Nellist 2011
Local Structure, Defects	HRTEM	Williams and Carter 1996
	HAADF-STEM, ABF-STEM	Scherzer 1936; Williams and Carter 1996; Loretto and Smallman 1975; Van Tendeloo et al. 2004
	Dark-field/bright-field TEM (DF/BF TEM)	Kossel and Möllenstedt 1939; Williams and Carter 1996; Van Tendeloo et al. 2004
	Weak beam imaging	
Defects and Strain	DF holography	Vòlkl, Allard and Joy 1998; Carter and Williams 2016
	CBED	Hÿtch, Snoeck and Kilaas 1998; Toda, Ikarashi and Ono 2000; Stuer et al. 2001; Chuvilin and Kaiser 2005; Hÿtch and Houdellier 2006
	Geometric phase analysis (GPA)	
Morphology, Size, Distribution	BF TEM	Carter and Williams 2016
	HAADF-STEM	
Surface Structure of the Crystals	Reflection electron microscopy (REM)	Halliday and Newman 1960; Nielsen and Cowley, 1976

Chemical Composition	Energy-dispersive x-ray spectroscopy (EDXS)	Chandler 1977
	EELS mapping, energy-filtered TEM (EFTEM) HAADF-STEM	Carter and Williams 2016
Chemical State (Oxidation State, Coordination, Bonding)	EELS, ELNES	Egerton 1996; Ahn 2004; Egerton 2009; Tan et al. 2012; Turner et al. 2012
Optical Properties, Plasmon/ Phonon Mapping	EELS, STEM-EELS	Nelayah et al. 2007; Krivanek et al. 2014; Lagos et al. 2017
Magnetic Structure	Electron holography	Haine and Mulvey 1952; Cowley 1992b; Lichte 1993;
	Lorenz microscopy	Fuller and Hale 1960; Chapman 1984
	Differential phase contrast	Lopatin et al. 2016
3D Structure	Electron tomography	Frank 1992; Weyland and Midgley 2007; Friedrich et al. 2009; Fultz and Howe 2013
	STEM sectioning	Borisevich et al. 2006
	Single particle reconstruction	Van Aert et al. 2011
Quasi Equilibrium and Ultrafast Processes	In situ EM (liquid, electrochemical, heating, cooling, biasing, straining)	Banhart 2008
	Time-resolved ultrafast EM (UEM)	Zewail and Thomas 2010; Flannigan and Zewail 2012
Image/Diffraction Simulation EFTEM Simulation	Multislice, Bloch waves	MacTempasX software; JEMS software; Koch 2002; Rosenauer and Schowalter 2008; Verbeeck, Schattschneider and Rosenauer 2009; Kirkland 2010

It is now imperative that an electron microscope has a high brightness electron source, such as a field emission gun (FEG); a corrective system for spherical aberration (using an aberration corrector); and can keep chromatic aberration low by using a monochromator to produce a highly monochromatic beam with a probe size around 0.5 Å and large probe current. Such a microscope is able to reach the image resolution of 0.5 Å in TEM mode (Kisielowski, Erni and Freitag 2008) and approach the same resolution in scanning TEM (STEM) mode (Erni et al. 2009). Simultaneous advances in the stability of electronics, specimen holders (piezo stages), efficiency of detectors, and computer speed enable new opportunities for materials science. Another main advantage of aberration-corrected microscopes is the ability to work at low voltages (80 kV and lower), bellow the knock-on damage threshold of many beam-sensitive materials, like carbon nanotubes, graphene, and metal organic framework (MOF) materials (Wiktor et al. 2017), and still have resolution around 1.36 Å for 80 kV.

In STEM mode, the image forms by scanned e-beam formed fine probe (~1 Å). Depending on the transmitted electrons that are used to form the image, an annular bright-field STEM (ABF-STEM) or high-angle annular dark-field STEM (HAADF-STEM) can be formed simultaneously within the same scanning. HAADF-STEM imaging is also known as Z-contrast imaging because the Rutherford-like scattering of the electrons elastically scattered to high angles leads to the image contrast, which is approximately proportional to the thickness of the crystal and the square of atomic number Z ($I_{HAADF\text{-}STEM} \sim Z^2$). Therefore, the image formed in HAADF-STEM is incoherent and is no longer a phase contrast image like in HRTEM and can be directly interpreted in terms of mass-thickness projection of the structure on the plane to show where the atoms are.

By using an annular detector to collect the electrons scattered by the sample at large angles, HAADF-STEM image and collection of forward scattered electrons using the GIF energy filter can be simultaneously realized. Analyses of collected inelastically scattered electrons using an electron energy-loss spectrometer (EELS) allow the determination of the electronic structure (the bonding and electronic state), chemical composition of the sample, and other sample properties. This makes the HAADF-STEM-EELS combination unique and powerful for direct imaging of the structure and getting local chemical/electronic information down to the sub-angstrom level with high-energy and spatial resolution. Moreover, relatively simple direct interpretation of the contrast in HAADF-STEM images opens the door for quantitative electron microscopy, helping to detect and quantitatively measure all the signals collected from a sample.

In almost all modern microscopes, even without aberration correctors, the classical basic set for characterization of materials turns out to be the ED, HAADF (ABF)-STEM, and simultaneously acquired EDX/EELS spectroscopy. Such combination of techniques does not require any special equipment and, as described from the Table 3.1, provides all necessary structural and chemical information for a material scientist. Indeed, HRTEM still can be used for a number of simple cases of instability of materials, but, for more complex materials such as thin films and functional materials with low levels of doping and nanomaterials, HAADF (ABF)-STEM is absolutely necessary.

This chapter aims to present some practical examples of using the "basic" set of advanced TEM and the role of each method. We show the discovery of new structures, which became possible only due to advancements of modern TEM techniques.

3.2 METHOD TO CHARACTERIZE MATERIALS BY MEANS OF ELECTRON MICROSCOPY: CASE STUDIES

3.2.1 The Manganoferrite $Pb_{2-x}Ba_xFeMnO_5$

Reported almost 20 years ago, the new $Pb_{1.33}Sr_{0.67}Fe_2O_5$ compound (Raynova-Schwarten, Massa and Babel 1997), with a structure close to a perovskite, has stimulated researchers to synthesize a series of ferrites closely related to this perovskite, with a composition similar to that of the brownmillerite family, $Ca_2Fe_2O_5$ or $Sr_2Fe_2O_5$ (Bertaut, Blum and Sagnieres 1959; Greaves et al. 1975). The brownmillerite structure is built up with FeO_5 tetragonal pyramids interconnecting the FeO_6 octahedra instead of the FeO_4 tetrahedra. Bearing in mind that trivalent manganese is able to occupy positions in octahedral and pyramidal sites, it is consequentially a potential candidate to generate structures similar to the ferrite brownmillerite. To understand the different behavior of manganese with respect to iron in this structural family, the substitution of Mn^{3+} for Fe^{3+} in the oxide $Pb_{1.33}Ba_{0.67}Fe_2O_5$ has been attempted. Using neutron powder diffraction (NPD), it was found that up to 50% of the Fe sites can be occupied in the final structure, resulting in a compound with a nominal formula: $Pb_{1.26}Ba_{0.74}MnFeO_5$. Room temperature XRD revealed that the structure of this compound is as an orthorhombic unit cell with space group *Pnma* and unit cell parameters $a = 5.7914$ Å, $b = 3.8888$ Å, and $c = 21.4064$ Å. However, no distinction between Fe and Mn atoms is available from XRD data because of the similarity in scattering power of Mn and Fe. Therefore, one of the main questions to be addressed via TEM analysis is the presence of Fe and Mn, its localization in the structure, and possible Fe-Mn interaction. (Barrier et al. 2013)

Figure 3.1 shows the results of typical TEM analysis for $Pb_{1.26}Ba_{0.74}MnFeO_5$. The ED patterns of $Pb_{1.26}Ba_{0.74}MnFeO_5$ were obtained for the four most relevant zone axes: [001]*, [010]*, [100]*, and [110]* (Figure 3.1a) and were completely indexed based on an orthorhombic *Pnma* structure obtained from XRD. No superstructure spots were detected in the ED patterns, possibly due to the Fe-Mn ordering. The shape of the diffraction spots and absence of streaks suggest a perfect crystal structure, free of extended defects.

The BF HRTEM image of $Pb_{1.26}Ba_{0.74}MnFeO_5$ was performed along the most informative [010] zone axis. (Figure 3.1b, left) and confirmed perfect crystallinity. However, addressing/solving the real atomic arrangement of the atoms in such a complex oxide structure from HRTEM is quite difficult, even using classical image simulation because of the lack of information obtained from XRD. In this respect, HAADF-STEM has been used and offers detailed chemical information. The HAADF-STEM imaging actually duplicates the HRTEM image; the structure can be qualitatively derived from the HAADF-STEM image. In the HAADF-STEM image (Figure 3.1b, right), where the contrast is directly proportional to the atomic number ($\sim Z^2$), the brightest dots correspond to the heaviest Pb ($Z = 82$) atoms, whereas less

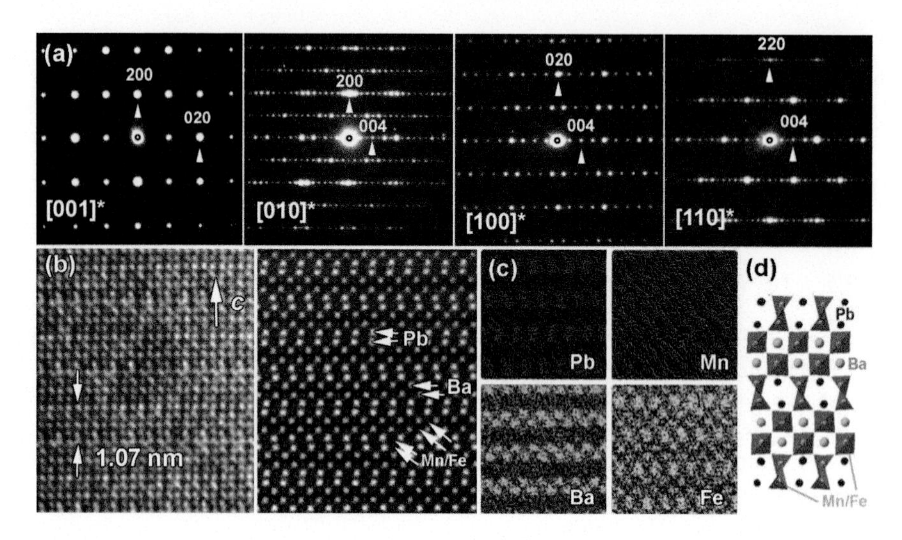

FIGURE 3.1 (a) ED patterns of $Pb_{1.26}Ba_{0.74}MnFeO_5$ obtained for four main zone axes: [001], [010], [100], and [110]. (b) Left panel: BF HRTEM for the structure in the [010] zone axis and parallel acquired HAADF-STEM image (right panel). (c) The corresponding atomic EDX elemental mapping for all elements: Pb, Mn, Ba, and Fe. (d) The exact structural model derived from a combination of XRD, ED, HAADF-STEM, and EDXS mapping. TEM data: Pb is red, Ba is blue, and Mn/Fe are orange. The corresponding position of all atoms in the HAADF-STEM image are depicted by white arrows.

bright dots are Ba (Z = 56) atoms. The weak bright dots in between the brightest dots correspond to the Fe (Z = 26) and/or Mn (Z = 25) columns. Thereby, the zigzag bright dots in the [010] HAADF-STEM image (Figure 3.1) indicate the positions of the perovskite layers, which consist of corner-sharing MO_6 octahedra (M = Fe/Mn), whereas Ba(Pb) cations sit in the tunnels of perovskite blocks (see overlay structural model in inset of Figure 3.1d).

Similarly, bright contrast dumbbells represent the pairs of Pb^{2+} cations (Pb_2) located inside each six-sided tunnel built up of edge-sharing MO_5 tetragonal pyramids, imaged as weak bright dots of Fe(Mn) atomic columns, and connected to perovskite blocks. Note, the "Pb_2" pairs are not parallel along the *c*-axis, but they are alternatively tilted left and right with respect to the *c*-axis, in agreement with the structure model determined from NPD data. Based on ED studies and HAAD-STEM images showing a uniform contrast free of any modulations, it is reasonable to conclude a good crystallinity of the material and a regular layer stacking along the *c*-axis. However, due to the small difference in the atomic number of Fe and Mn, it is almost impossible to distinguish these atoms directly from the HAADF-STEM image. It was shown that the difference in image contrast of individual atomic columns in HAADF-STEM is possible only if the difference in the averaged Z number is larger than 3 (Van Aert et al. 2009). In this respect, atomic EDX elemental mapping was used. Because the EDX elemental mapping and the HAADF-STEM image acquisition can be performed simultaneously, direct chemical and structural correlation can be obtained. Moreover, the quantitative information on cation ratio also can be obtained from the same measurements and give $Pb_{1.26}Ba_{0.74}MnFeO_5$, which is close to expected nominal composition.

Figure 3.1c shows a chemical mapping for $Pb_{1.26}Ba_{0.74}MnFeO_5$ along the [010] zone axis for all the elements. It was found from the Fe and Mn atomic maps that these elements are distributed randomly within the same octahedral and pyramidal layers: there is no preferential position for Fe or Mn and NO ordering or segregations have been observed, which is in good agreement with ED studies. It should be noticed that the appearance of the weak contrast dots in the positions of the Ba atoms in the Pb map suggests the presence of small amounts of Pb atoms (around 10%) in the perovskite cages. The intermixing of the Pb^{2+} and Ba^{2+} cations in crystallographic distinct position in perovskite cages was later confirmed by NPD study (Barrier et al. 2013).

Using advantages of TEM to visualize the structure of materials at atomic column level within several atomic layers, HRTEM observation of different samples of this material along the [010] zone axis has revealed the presence of extended defects running along *a*-axis, as shown in Figure 3.2. Note that these defects are difficult to detect in the BF HRTEM image (Figure 3.2a), whereas the high-resolution HAADF-STEM image (Figure 3.2b) emphasizes the structure of these defects, showcasing the advantage of using the HAADF-STEM technique to characterize for complex structures.

As mentioned above, the perfect structure (Figure 3.1d) consists of Pb_2-(001) successive layers, which are alternately oriented left (L) and right (R) along the *c*-axis, forming a sequence of type "L-R-L-R...". Indeed, the largest part of the image (Figure 3.2b), showing the structure along [010] zone axis, exhibits this sequence. However, along with this regular L-R-L-R... sequence, one observes stacking faults along the *c*-axis that consist of two similar oriented layers R and R next to each other. This stacking fault defect can be described as ...- R-L-R-R-L-R-... and is represented in the structural model in Figure 3.2c. Note that the ED patterns did not reveal any extra spots along [001] in any crystallite, suggesting that no long-range ordering of such extended defects takes place in the matrix. The absence of any intensity modulations confirms that the defect is a purely structural defect and has no chemical substitution nature.

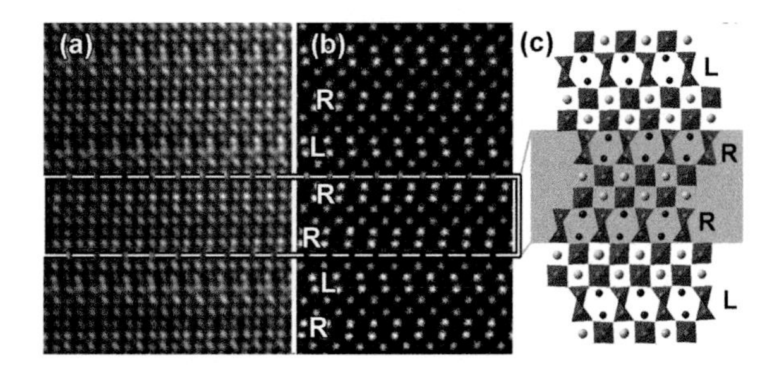

FIGURE 3.2 (a) BR [010] HRTEM and (b) HAADF-STEM images of the stacking fault region in $Pb_{1.26}Ba_{0.74}MnFeO_5$. (c) Corresponding structural model. The stacking fault region is marked with a dashed rectangle and correspondingly by gray area in the structural model. L, Left; R, right.

The combination of ED studies, HAADF-STEM imaging, and EDX analysis is powerful, easily accessible, and an essential set of microscopy techniques that allow one to fully characterize complex materials and to obtain structural and chemical information about a sample. Among the numerous similar techniques, EDX has obvious advantages, such as requiring minimal setup, a noncritical dependency to thickness of the sample, easy and reliable quantitative analysis, almost no Z number limitation, and a high signal-to-background ratio. Through all these advantages, EDX has some limitations, which included difficulty in sensing light elements with a small Z number, nonlocal fluorescence, and limited thin film signal-to noise ratio. In this respect, EELS is complementary to the EDX technique and can be used for analysis of both chemical composition and electronic structure of materials. Using both techniques helps in solving the problem of certain overlapping signals, which might occur for some elements

3.2.2 Rare-Earth Cations Inside BaLaCuP Clathrate Polyhedral Cages Revealed by Advanced TEM

The attempt to incorporate the trivalent rare-earth cations inside the oversized polyhedral cages of three-dimensional framework has a long history and was realized recently in $Ba_{8-x}R_xCu_{16}P_{30}$ clathrates (Wang et al. 2018). The synchrotron powder XRD pattern of $Ba_{6.4}La_{1.6}Cu_{16}P_{30}$ is shown in Figure 3.3 (upper panel).

All the intense diffraction peaks can be indexed in the primitive tetragonal cell, $a = 14.0655(7)$ Å and $c = 10.0037(6)$ Å. No substantial alteration of the Cu/P ratio in the framework was observed by electron microprobe analysis, and the cation ratio was determined as $Ba_{7.1(1)}La_{0.9(1)}$. It is obvious that XRD patterns of $Ba_{8-x}R_xCu_{16}P_{30}$ show a systematic peak shift to the higher diffraction angles with the rare-earth content increase indicating a reduction of the unit cell volume (Figure 3.3, insert in upper panel). However, XRD is not suitable to distinguish between La and Ba atoms due to similar x-ray scattering factors for these elements. One of the main questions addressed to thew TEM was where La and Ba atoms are situated inside the Cu/P framework.

The ED patterns collected along three main zone axes are shown in Figure 3.3 and confirmed the tetragonal structure. All ED patterns can be completely indexed in the $P4_2/ncm$ space group with $a = 14.06$ Å and $c = 10.00$ Å unit cell parameters. However, the [100] ED pattern (Figure 3.3, rightmost panel) exhibit very weak superstructure 001 spots along [001] directions that are not detectable by XRD. High-resolution HAADF-STEM images (Figure 3.4), obtained along main crystallographic zones [001], [011], and [100], are in good agreement with simulated images and the structural models (Figure 3.4, right column). The brightest contrast dots in the HAADF-STEM images correspond to Ba or La (Z = 56 and 57) atomic columns, and the less bright dots correspond to Cu atomic columns (Z = 29), whereas P atoms (Z = 15) are difficult to visualize in HAADF-STEM due to low relative Ba/La brightness. Careful analysis of high-resolution [100] HAADF-STEM images revealed the reason for the superstructure spots appearance: they are caused by the intensity modulations in the neighboring Ba/La and Cu/P columns.

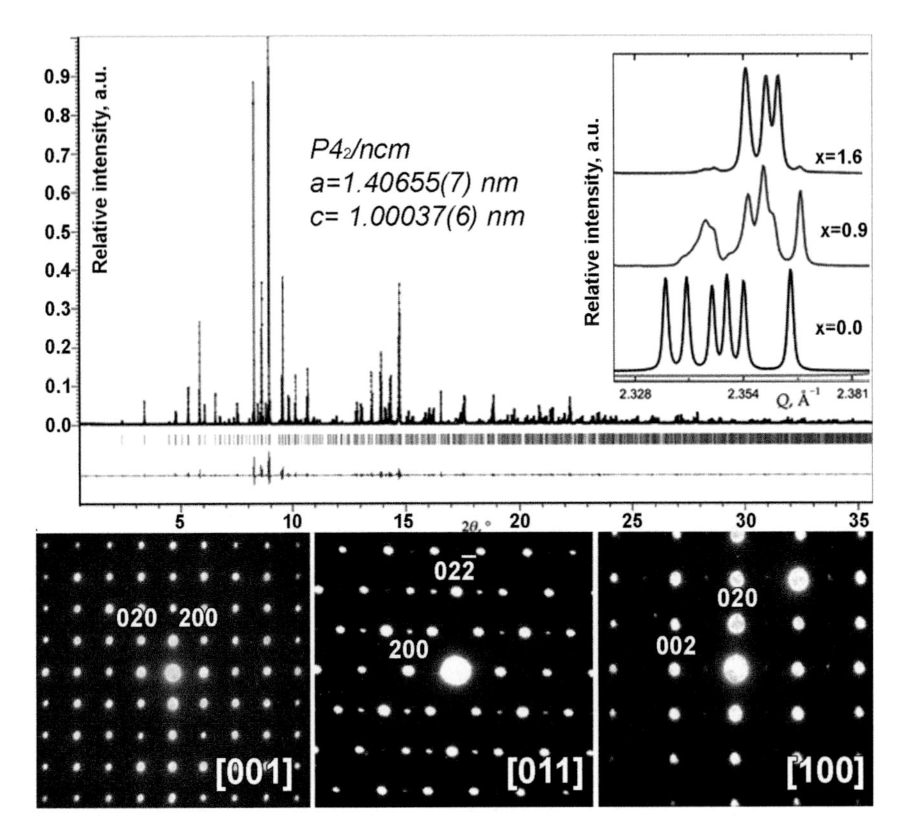

FIGURE 3.3 Upper panel: Synchrotron powder XRD pattern of $Ba_{6.4}La_{1.6}Cu_{16}P_{30}$ and fragments of synchrotron powder diffraction patterns for $Ba_{8-x}La_xCu_{16}P_{30}$ (x = 0, 0.9, and 1.6). Bottom panel: ED patterns for the three most relevant zone axes of tetragonal $Ba_{6.4}La_{1.6}Cu_{16}P_{30}$ clathrate: [001], [011], and [100].

These are visible especially in the appearance of a relatively bright single dot among six Cu/P atomic columns in the HAADF-STEM image. Bearing in mind that $I_{HAADF\text{-}STEM}{\sim}Z^2$ and taking into account the structural model of $Ba_{6.4}La_{1.6}Cu_{16}P_{30}$ and simulated images (Figure 3.4), it can be concluded that these relatively bright dots correspond to a Cu-rich atomic column. No difference between La (Z = 57) and Ba (Z = 56) column contrast can be detected due to a close Z number. Unfortunately, simple and reliable EDXS atomic elemental mapping has little use for this particular structure due to nearly overlapping characteristic peaks (Ba, L = 4.465 keV; La, L = 4.650 keV). The position and the presence of La in the crystal structure of $Ba_{6.4}La_{1.6}Cu_{16}P_{30}$ was confirmed by atomic resolution EELS-STEM (Figure 3.5). The elemental EELS mapping of $Ba_{6.4}La_{1.6}Cu_{16}P_{30}$ along the [001] and [011] zone axis clearly demonstrated the location of La and Ba in crystallographic distinct positions: La atoms are located inside the smaller pentagonal dodecahedra, whereas larger Ba atoms are located inside tetrakaidecahedra. Moreover, as seen in Figure 3.5, the EELS atomic resolution image along the [100] zone axis

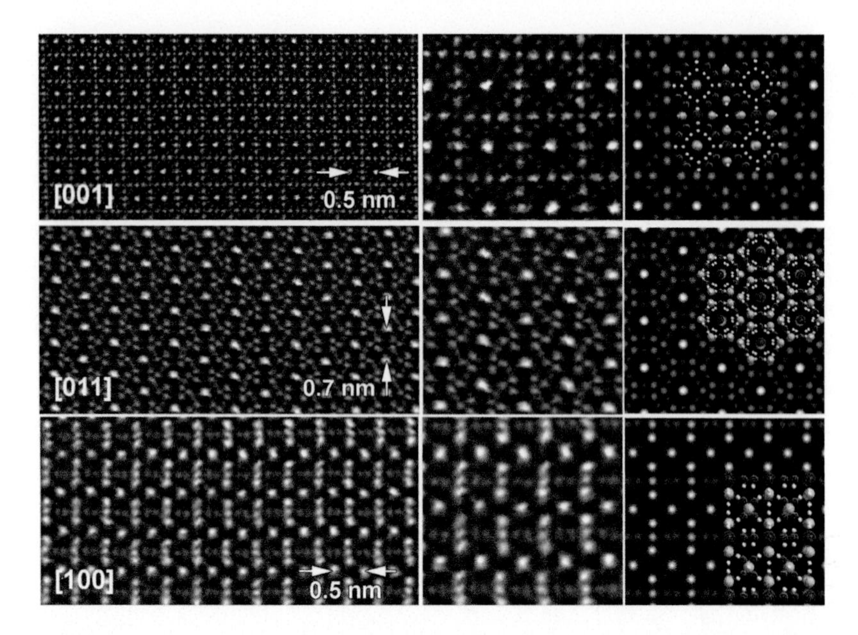

FIGURE 3.4 High-resolution HAADF-STEM images for the three most relevant zone axes of tetragonal $Ba_{6.4}La_{1.6}Cu_{16}P_{30}$ clathrate: [001], [011], and [100]. The corresponding enlargement experimental (middle) and simulated HAADF-STEM images together with overlay structural models are in the right panel with the following color coding: La is blue, Ba is green, Cu is red, and P is yellow.

unambiguously confirms a superstructure ordering model. The Cu-rich columns (red in Figure 3.5 color overlay, bottom panel) are located exactly opposite to La atomic columns (green in Figure 3.5 color overlay, bottom panel).

In conclusion, the obtained EELS data allow us to localize the position of La and Ba inside the clathrate cages. They are in good agreement with synchrotron x-ray and neutron diffraction data, and they confirm the proposed structural model. Our detailed structural analyses unambiguously show that La was indeed incorporated to the small cages of Cu-P clathrate I framework.

Finally, the combination of atomic resolution EELS and STEM studies with synchrotron XRD and NPDs unambiguously demonstrated the residence of rare-earth elements in the clathrate cages (Wang et al. 2018).

3.2.3 CuInSe Hexagonal Flat Nanoparticles

A ternary $CuInSe_2$ semiconductor having chalcopyrite structure is an outstanding material for several important optoelectronic applications, such as photovoltaic (PV) devices (Guo et al. 2008; Jeong et al. 2012), light-emitting diodes (LEDs) (Zhong et al. 2011), and photocatalytic H_2 generation (Sheng et al. 2014), due to a high absorption coefficient of $\approx 10^{-5}$ cm^{-1} and a direct band gap of 1.04 eV (Zhong et al. 2011). Recently, it was shown that chalcopyrite $Cu(In,Ga)Se_2$ used as a material for

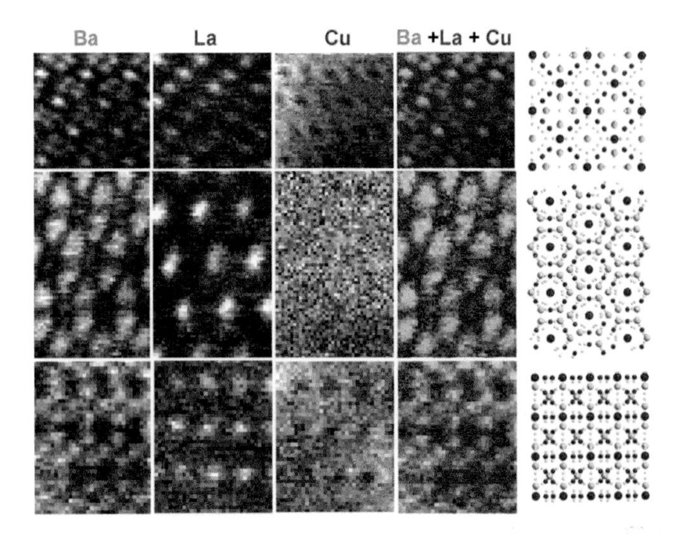

FIGURE 3.5 Atomic resolution EELS elemental mapping for the [001], [011], and [100] zones of $Ba_{6.4}La_{1.6}Cu_{16}P_{30}$: Ba M5,4 map, La M5,4 map, Cu L3. Color overlays: Ba in green, La in blue, and Cu in red, together with a structural model (correspondingly, La is blue, Ba is green, Cu is red, and P is yellow). Right panel: Structural model based on corresponding EELS elemental mapping similar to EELS mapping color code for atoms.

thin-film solar cells reached the record efficiency of 22.8% (Kamada et al. 2016). However, PV cell fabrication requests high temperature and high vacuum deposition techniques. At the same time, solution-based printing technologies offer more effective thin-film PV fabrication (Guo et al. 2008). Therefore, one challenge will be to develop the synthesis of $CuInSe_2$ (NPs) for ink formulation, which later can be used in the screen printing of $CuInSe_2$-based PVs. They can be solution processed to obtain the photoabsorber layer. It is important to notice that the optical and electronic properties of such devices strongly depend of the size, morphology, and structure of the NPs (Yarema et al. 2013). The synthesis method using colloidal sample preparation techniques for $CuInSe_2$ NPs results in NPs with chalcopyrite type crystal structures or in a mixture of different phases (Sheng et al. 2014), and very limited reports on synthesis wurtzite structures and influence on PV performance. Therefore, one of the main goals of this work was to obtain single-phase $CuInSe_2$ NPs with a wurtzite hexagonal structure and preferentially platelike morphology appearance using colloidal synthesis (Sousa et al. 2019).

TEM results on $CuInSe_2$ NPs are displayed in Figure 3.6. In Figure 3.6a, $CuInSe_2$ NPs obtained by colloidal synthesis imaged using HAADF-STEM and ED patterns (Figure 3.6b) obtained from two representative zone axes $[001]_h$ and $[100]_h$ are collected. All ED patterns can be indexed based on the wurtzite hexagonal $P6_3mc$ structure using the following cell parameters obtained from XRD (Figure 3.6a): $a = 4.08$ Å and $c = c_{wurtzite} = 6.69$ Å. However, ED patterns shown in Figure 3.6c clearly indicate the doubling of *at least* one unit cell dimension with respect to the XRD data

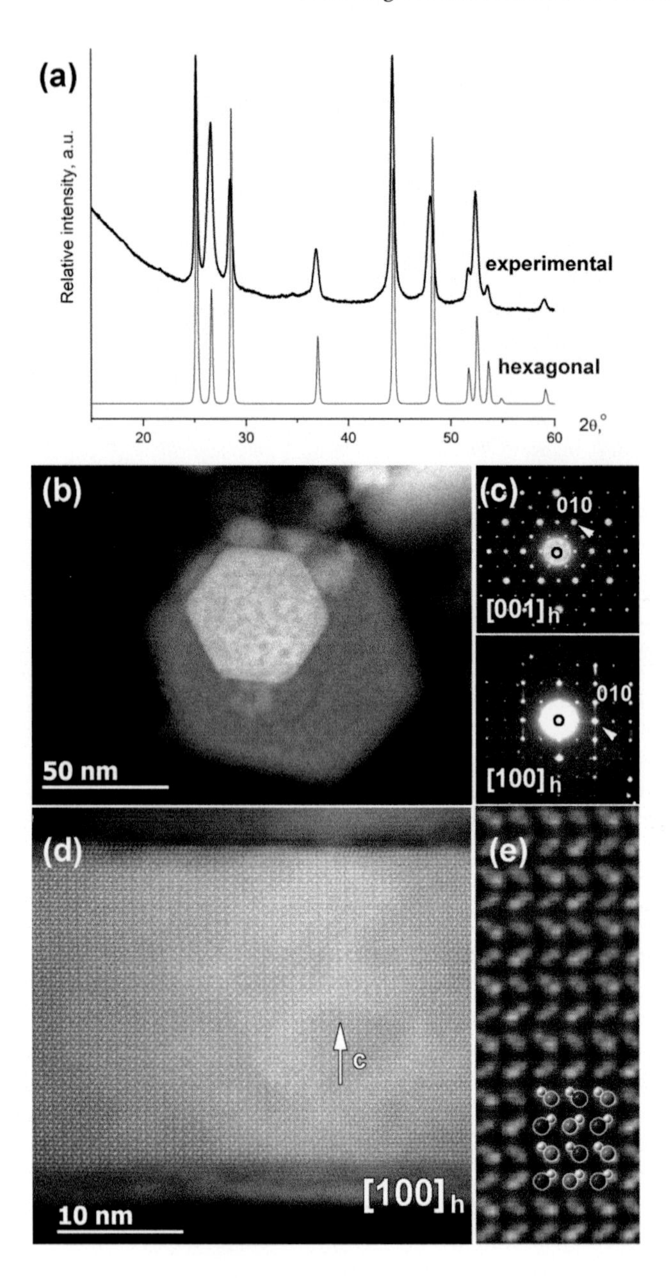

FIGURE 3.6 (a) Experimental XRD pattern of the $CuInSe_2$ synthesized NPs and calculated patterns for the disordered hexagonal $P6_3mc$ subcell with one Cu/In position. (b) Low-magnification HAADF-STEM image of hexagonally shaped CuInSe NPs. (c) ED patterns along two representative zone axes: $[001]_h$ [(b) top panel] and $[100]_h$ [(d), bottom panel] can be indexed based on a hexagonal structure. (e) High-resolution $[100]_h$ HAADF-STEM image of hexagonally shaped NP viewed along the *c*-axis. (d) Magnified image with overlapped structural model, in which there are large dark atoms, large gray-colored atoms correspond to Cu, and small bright atoms are Se.

($a = 2a_{wurtzite}$ = 8.164.08 Å and $c = c_{wurtzite}$ = 6.69 Å). It is well known that the shape of NPs most often approximately reflects the structure/symmetry of the NP material (Wu, Yang and Wu 2016). Figure 3.6a shows representative low-magnification HAADF-STEM images of basal- and prism-facet views. The particles are highly crystalline and exhibit a hexagonal shape that is fully consistent with the hexagonal structure determined by ED. No segregation or secondary phases were detected by EDX elemental mapping, indicating the phase purity of the product and homogeneous distribution of all elements within a single NP. The high-resolution HAADF-STEM image along the *c*-axis in Figure 3.6c depicts the amorphous and the partially crystallized layers on the surface of the $CuInSe_2$ NPs, which most likely originate from partial surface oxidation and/or surface deficiency in cations, which is typical for nonoxide NPs. The structural model of hexagonal $CuInSe_2$ is overlapped on the high-resolution HAADF-STEM image (Figure 3.6d) and demonstrates good agreement with an experimental image. However, careful high-resolution HAADF-STEM studies along $[001]_h$ zone axes reveal important information about the atomic level arrangements in the $CuInSe_2$ NPs, which cannot be extracted from CTEM and XRD data.

A careful inspection of the high-resolution [001] HAADF-STEM images shown in Figure 3.7 reveals an unambiguous contrast variation along the [110] direction with alternating bright and dark atomic columns pairs (Figure 3.7c), which is a signature of the different chemical compositions of these atomic columns. This suggests a local chemical ordering. The FT pattern obtained from ordering region also exhibits a doubling of the unit cell parameter, similar to that observed in [001] ED patterns. Such ordering is inconsistent with hexagonal or trigonal symmetry and has never been reported before. Bearing in mind a high similarity of the ionic radii for Cu^+ and In^{3+} in tetrahedral coordination, 0.60 and 0.62 Å, respectively, and the chemical composition obtained from EDX elemental mapping ($CuInSe_2$), it was suggested that the Se sublattice is arranged in a perfect hexagonal framework identical to those in the wurtzite subcell, and Cu and In atoms are ordered over distinct framework positions resulting in observed contrast. The Se framework is responsible for the overall hexagonal morphology appearance of the particles and ED patterns. Moreover, careful analysis of high-resolution HAADF-STEM images revealed chemical twinning of the ordering domains. An intergrowth of the differently ordered 60-degree rotated twinning domains is visible in the HAADF-STEM images (Figure 3.7a,b). The chemical ordered Cu/In domains are twinned within the hexagonal arranged Se framework (Figure 3.6d model). This explains the hexagonal appearance of the particle morphology and the overall hexagonal symmetry in [001] ED diffraction patterns. Founded Cu/In ordering requests the reduction of the structure symmetry to *Pm*. An orthorhombic model in $Pmc2_1$ (a = 4.0863 Å, b = 7.0819 Å, c = 6.7405 Å, $\alpha = \beta = \gamma = 90°$) suggested by the computational work of Lau et al. tackling different Cu/In ordered models of wurtzite $CuInSe_2$ (Xu et al. 2012) was considered. Based on TEM results, the XRD pattern was revised and the careful examination of the pattern reveals several low-intensity reflections that cannot be assigned to the tetragonal chalcopyrite $CuInSe_2$ or any other possible admixture phases. It perfectly fits with an orthorhombic model in $Pmc2_1$.

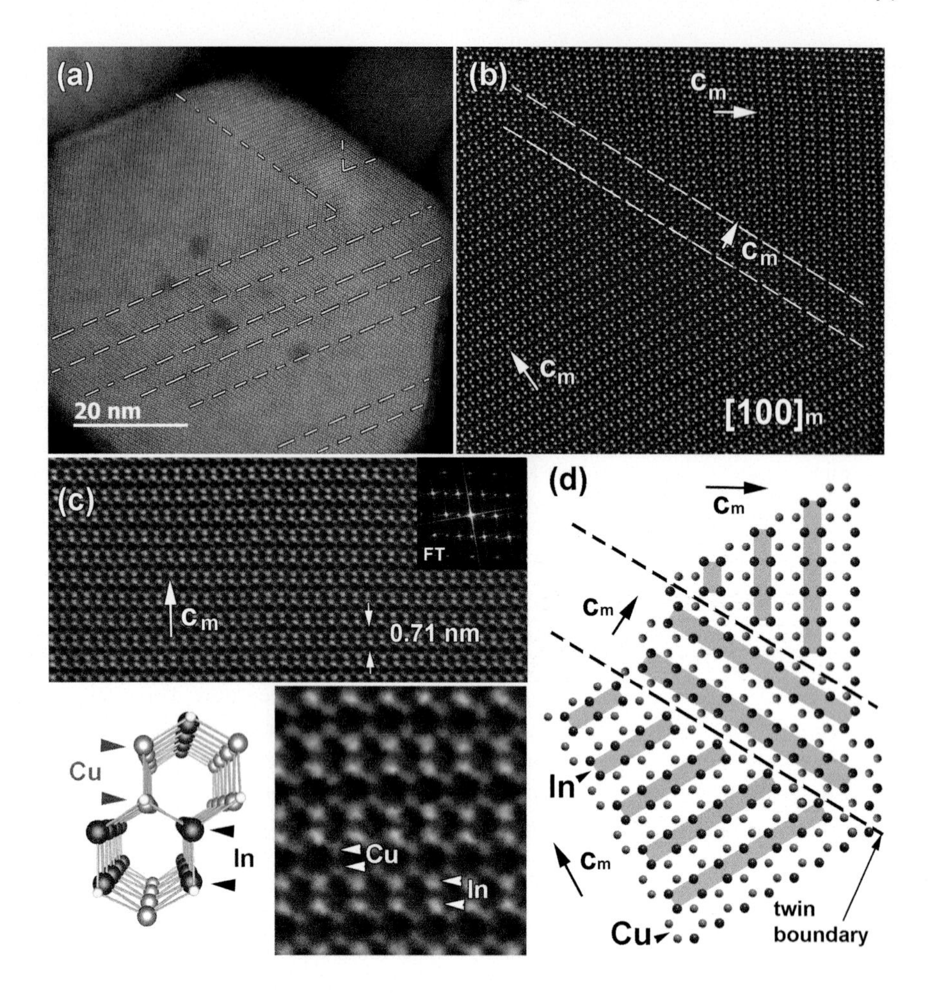

FIGURE 3.7 (a) HAADF-STEM image of hexagonal shaped $CuInSe_2$ NP. (b) High-resolution HAADF-STEM image of the NP in (a) showing a twinning monoclinic structure area. (c) Selected high-resolution HAADF-STEM image with corresponding FT pattern along the [100]m zone axis demonstrated In/Cu resulting in *Pm* monoclinic structure. The position of In (bright dots) and Cu (less bright dots) indicated in magnified image (insert, right bottom corner). The corresponding structural model of In-Cu ordering is shown in the left bottom corner. (d) Corresponding to (c) schematic drawing of twinning area where In atoms are black and Cu atoms are gray. The ordered In layers are marked by gray background stripes. Twins boundaries are marked by dashed lines in all figures (a, b, and d).

3.2.4 Revisiting Hollandite $Bi_{2-x}V_yV_8O_{16}$ Cubic Structure Thanks to HAADF-STEM

Due to the number of interesting physical properties related to electronic correlations, magnetic frustration, and orbital ordering in hollandite structures with the general formula $A_xM_8O_{16}$ (A = K, Pb, Ba, Bi, Ag,…,M are magnetic transition metals with mixed valence), scientific interest has grown over the past years. The crystal structure of $Bi_{1.8}V_8O_{16}$ consists of stripes of stacked edge-sharing VO_6 octahedra, which share corners to form a V_8O_{16} framework with wide and narrow tunnels. The larger tunnels are occupied by Bi^{3+} cations. The charge transfer from the A^{n+} cations to the V_8O_{16} hollandite network generates a V^{3+}/V^{4+} mixed valence and the extra electrons ordering in a one-dimensional (1D) pattern appear, which is leading to complex V-V magnetic exchange pathways. Through corner-sharing oxygen atoms bridging the stripes, there exists an additional magnetic exchange between the zigzag chains along the V_8O_{16} stripes. XRD of $Bi_{1.8}V_8O_{16}$ confirms the hollandite tetragonal structure with the space group *I*4/*m* previously reported for a $Bi_{1.625}V_8O_{16}$ single crystal (Abraham and Mentre 1994) and unit cell parameters were found to be $a = 9.9323(2)$ Å and $c = 2.9128(1)$ Å ($V = 287.35(1)$ Å^3) (Lebedev et al. 2017).

ED patterns collected along the three representative [001], [100], and [111] zone axes are shown in Figure 3.8 and were indexed based on the cubic *I*4/*m* structure using the unit cell parameters determined by powder XRD. No superstructure spots resulting from possible long-range ordering were observed. However, a careful inspection of [100] and [111] ED patterns revealed the appearance of weak diffuse streak lines parallel to the *h*0*l* spot row, which might be interpreted in terms of short-range ordering (Figure 3.8a). Moreover, diffraction spots in the [001] ED pattern are not round, but slightly elongated along the [531] direction, suggesting some disorder along this direction.

To explain these features we examined the ED and XRD patterns. The overall structure should be consistent with the reported hollandite structure, and all structural changes occur at nanoscale. Here, the HAADF-STEM imaging was used to clarify the crystal structure model of $Bi_xV_8O_{16}$ and determine possible structural and/or compositional variations at the atomic level. Due to the large difference in atomic number between Bi ($Z = 83$) and V ($Z = 23$), the columns of Bi and V atoms can be easily distinguished, making it possible to detect even small changes in composition.

A high-resolution HAADF-STEM and simultaneously acquired ABF-STEM images of the $Bi_{1.8}V_8O_{16}$ structure along the main and most informative [001] zone axis are shown in Figure 3.7b. The two images are complementary. The Bi and V atomic columns are easily distinguishable in the HAADF-STEM image (Figure 3.8b): the brightest square arrangement dots correspond to the Bi atomic columns ($Z = 83$), whereas the eight less bright dots surrounding each Bi atomic column are V atomic columns ($Z = 23$). Figure 3.9 shows the HAADF-STEM image of $Bi_{1.8}V_8O_{16}$ taken along the orthogonal [100] direction. In the [100] HAADF-STEM image (Figure 3.9), the two rows of bright dots correspond to a row of Bi atoms and the less bright dots in between Bi are the V atoms.

The structure looks well crystalline and free of any extended defects. However, a careful inspection of the images reveals the presence of the vanadium columns, whereas the contrast in the vicinity of the Bi columns is strongly reduced in HAADF-STEM and ABF-STEM images (Figure 3.8b). This reduction or luck of the contrast can be attributed to the presence of stripes of Bi vacancy channels in $Bi_{1.8}V_8O_{16}$. HAADF-STEM imaging along the [100] zone axis also revealed the presence of Bi vacancies (Figure 3.9). According to the structural model (Figure 3.9, insert right bottom corner) the Bi columns should image in the [100] HAADF-STEM image

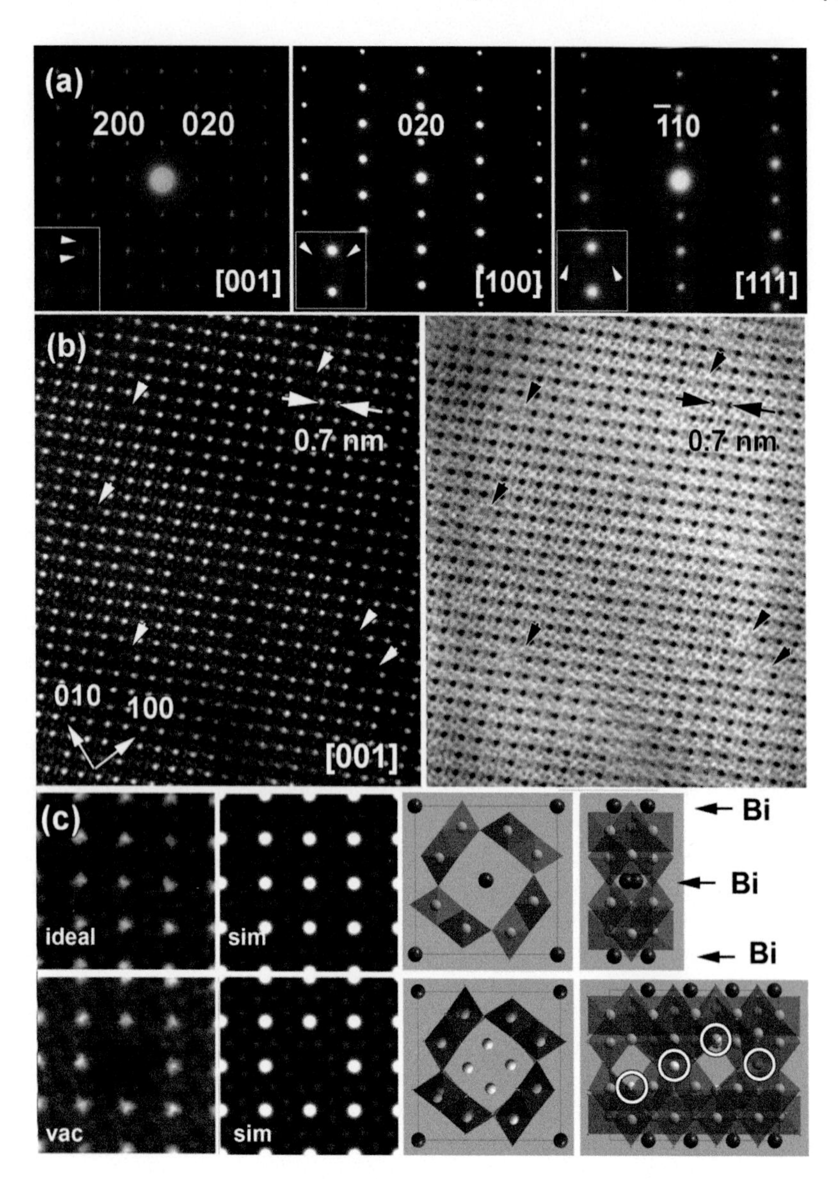

FIGURE 3.8 (a) ED patterns for three representative zone axes, [001], [100], and [111], confirming delafossite structure of BiVO. Note the elongation of diffraction spots in the [001] ED pattern and the presence of streak lines around 010 and 110 spot rows in [100] and [111] ED patterns, respectively. (b) High-resolution HAADF-STEM and simultaneously acquired ABF-STEM images of $Bi_{1.8}V_8O_{16}$ recorded along the [001] zone axis direction. (c) Magnified experimental image in (b) (left panel) and simulated [001] HAADF-STEM images (second from the left panel) together with a corresponding model of the ideal $Bi_{1.8}V_8O_{16}$ structure (upper right panel) and Bi vacancy column filled with square V cluster (bottom right panel). The models are given in two main projections: [001] (left column), and [100] (right column) along the *c*-axis showing the location of the V cluster along the vacancy channel (right bottom panel).

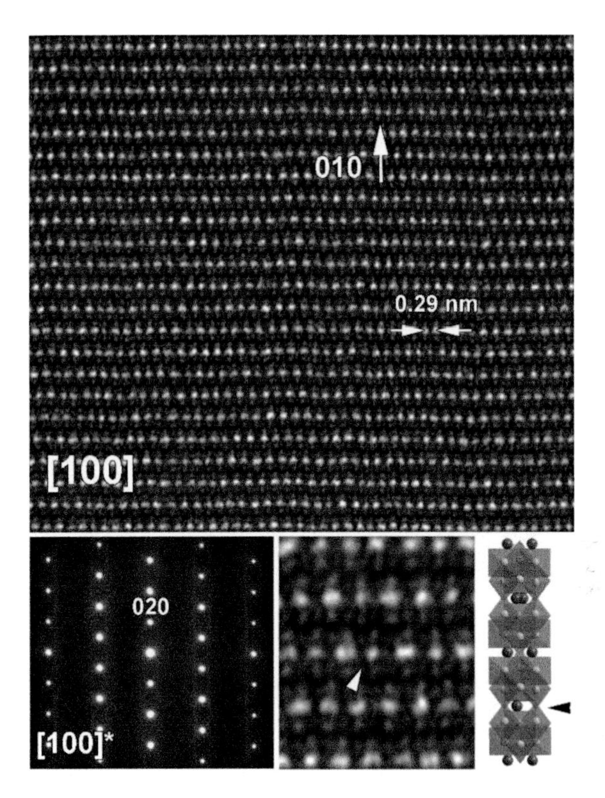

FIGURE 3.9 High-resolution [100] HAADF-STEM image of $Bi_{1.8}V_8O_{16}$ and corresponding ED pattern (left bottom panel). Notice the diffuse line along the *0hl* spot row. Magnified image and structural model are given at the right bottom corner. The columns with the Bi deficiency are depicted with a white arrowhead in the magnified image and correspondingly in the structural model by a black arrowhead.

as bright dumbbells. However, careful analysis of the [100] HAADF-STEM image (Figure 3.9) revealed numbers of dumbbells where, in spite of two Bi atoms, only one Bi atom can be observed (Figure 3.9, arrowheads). They are not ordered but randomly distributed over the crystallite structure. Moreover, the disappearance of the Bi columns within the wide channel is accompanied by the appearance of a square arrangement of lower contrast intensity dots displaying contrast similar to the other V columns, suggesting the presence of additional V cluster-like arrangements inside the empty wide channels (Figure 3.8c, bottom panel). Based on the obtained HAADF-STEM structural data on V location inside the empty wide channel and taking into account crystal chemistry consideration, a valuable structural model was proposed (Lebedev et al. 2017).

The structural model was constructed starting with the reported $P4_1$ crystal structure data of $Bi_{1.625}V_8O_{16}$ and unit cell parameters proposed from TEM data as $a = a_{Bi1.625V8O16}$ and $c = 4c_{Bi1.625V8O16}$. The atomic positions of the Bi, V, and O atoms of the surrounding the empty Bi channel structure were used without changes, as determined from the single crystal structure and the XRD experiment. Taking into account positions and interatomic distances determined from HAADF-STEM images, the following structural model of the Bi vacancy channel was proposed. A maximum intensity search was used to find the positions of the atomic columns of additional vanadium and adjacent oxygens. As it is evident from the high-resolution HAADF-STEM image (Figure 3.8b and c), the vanadium atoms are located within the empty Bi channel around the ½½*z*-axis, close to the wall of the channel employing the available oxygen atoms to form a tetrahedral environment and yielding the composition $Bi_1V_1V_8O_{16}$. The oxygen atoms were slightly shifted from their ideal positions in the crystal structure of $Bi_{1.625}V_8O_{16}$, keeping in mind reasonable V-O bond distances [around 2.0A; Krakowiak, Lundberg and Persson (2012)]. The stability of the proposed defect structure (Figure 3.8c) was examined using the first-principle calculations within the framework of the density functional theory (Lebedev et al. 2017). It was found that the defect configuration proposed in Figure 3.8c is stable, which was also confirmed by repeating the calculations with initial random displacement of all the atoms within 0.1 Å. The V atoms are located at different "heights," which breaks the symmetry and caused changes in the positions of neighboring Bi atoms. It should be noticed that another metastable configuration was found where V atoms were displaced by about 0.3 Å away from the center of the channel, but this configuration was higher in energy by 1.3 eV. HAADF-STEM image simulation (Figure 3.8c) using a DFT-optimized geometry structural model gives a good fitting with the experimental one and describes well the fine details of the experimental HAADF-STEM contrast.

The final composition in the case of such Bi-by-V substitution is $V_2V_8O_{16}$, i.e. $V_{10}O_{16}$. This would yield the average valence of $\nu_V = 3.2$, whereby the tetrahedral coordination of the additional position put into mind a possible presence of pentavalent vanadium that was confirmed by EELS. Also, the presence of extended defects supports the possibility of the pentavalent vanadium to be present. Figure 3.10a shows a low-magnification BF TEM image and corresponding ED pattern of a defect area. Diffraction spots appear as crosses and corresponding parallel bright lines in the TEM image create the so-called "parquet" type structure by rotation of lamellas by 90° with respect to each other. Such contrast is typically observed for a square network of misfit dislocations or might be due to periodical boundaries and/or a twin plane along certain direction. Interestingly, the direction of these dark contrast lines does not correspond to a low indexed direction as expected for a tetragonal structure (Lebedev et al. 2001) but is roughly perpendicular to a [531] direction. These lines also can be Moiré type fringes resulting from diffraction by two closely related sandwiched materials. From the [001] ED pattern (Figures 3.8a and 3.10a) the elongation of main diffraction spots along two orthogonal <531> directions is obvious. Figure 3.10b shows high-resolution HAADF-STEM images similar to the TEM image (Figure 3.10a) features; parallel dark lines of several nanometers' width are clearly visible. Their separation is in the range of 10 to 11 nm and the direction is roughly perpendicular to the [531] direction. Enlargement of the HAADF-STEM image (Figure 3.10c) revealed that the line contrast image results from a modulation of sharpness of the V atomic columns. No changes in the atomic positions and/or local structure difference from $Bi_{1.7}V_8O_{16}$ have been observed. A geometric phase analysis (GPA), which is a powerful method for strain mapping from HRTEM images

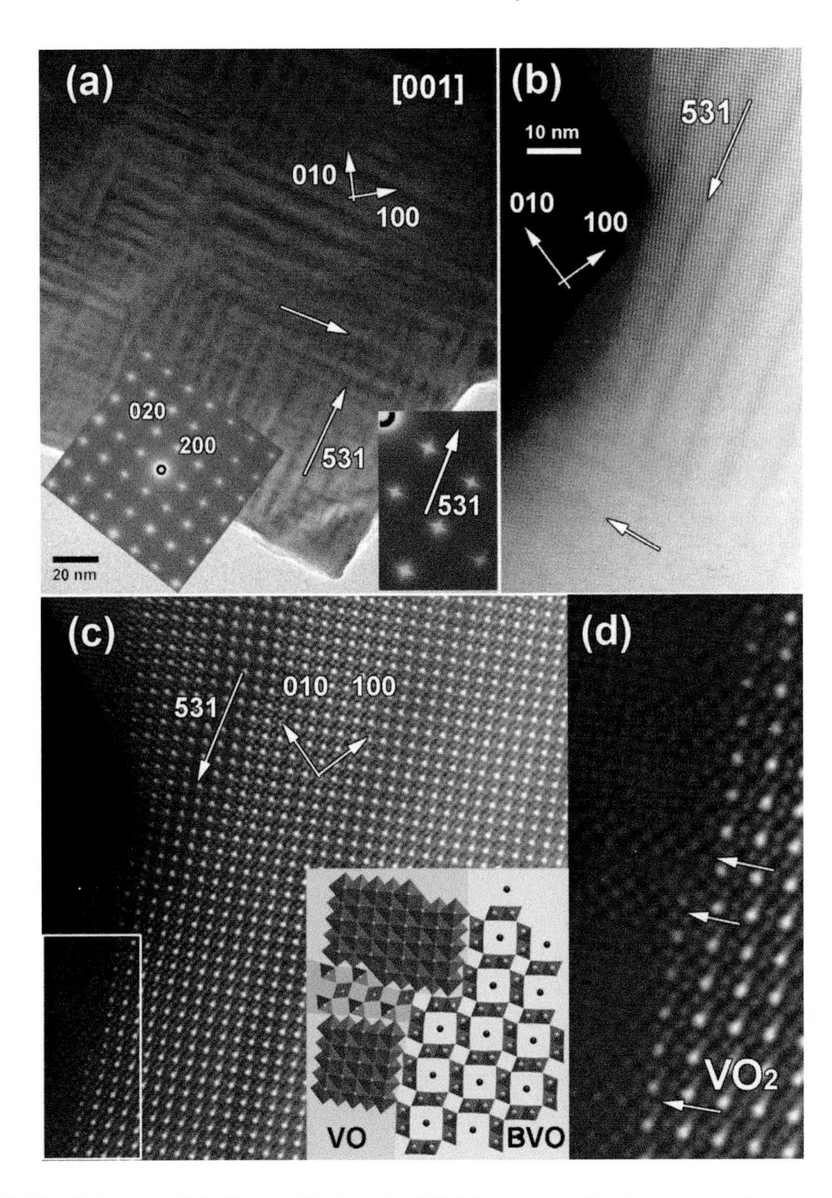

FIGURE 3.10 (a) BF TEM and (b) HAADF-STEM low-magnification images of type structure in $Bi_{1.8}V_8O_{16}$ and corresponding ED pattern given [(a), insert]. White arrows indicate twin boundaries. (c) High-resolution HAADF-STEM image of the crystal edge of the twinned structure and (d) magnified image showing an epitaxial intergrowth of VO and VO_2 structure within $Bi_{1.8}V_8O_{16}$. Corresponding structural model of intergrowth is given [(c), insert]. White arrows in (d) indicate nano-lamellas of VO_2 structure with an epitaxially intergrowth with VO and BVO structures.

(Hÿtch and Houdellier 2006), provides color maps of strain and shows pronounced color divergence at the boundary line across the boundary plane direction [531]. These data strongly suggest differences in the distances of two successive planes at the boundary with respect to the bulk material, which can be associated with a strain. The reason for such strain can be understood from a careful analysis of the HAADF-STEM image in Figure 3.10c. Enlargement of the edge of the crystallite is given in Figure 3.10d and clearly demonstrates three types of structure, which can be detected directly from the HAADF-STEM image. The first type is $Bi_{1.625}V_8O_{16}$ with a well-defined square arrangement of the bright contrast Bi atomic column and eight surrounding less bright V atoms. Two other structures are definitely free of Bi atoms and build up only from V and O atoms being VO and VO_2. Importantly, all three structures are epitaxially intergrown: the layer of VO and VO_2 is like a pancake within the $Bi_{1.7}V_8O_{16}$ crystal situated along [001]. The corresponding structural model is an insert in Figure 3.10c (bottom left panel). Bearing in mind that VO is an edge-sharing VO_6 octahedral structure (cubic space group $Fm\bar{3}m$, no. 225; $a = 4.063$ Å) and VO_2 is a vertex-sharing VO_6 octahedral structure (tetragonal space group $P4_2/mnm$, no. no. 136, $a = 4.54$ Å, $c = 2.88$ Å), the layer of VO-VO_2 will have some mismatch with the (001) plane of $Bi_{1.7}V_8O_{16}$ resulting in strain wave along the [531] direction. Therefore, strain contrast in the HAADF-STEM image of dark lines must be attributed to the curvature of the atom columns and most probably distortion of the VO_6 octahedral as most flexible, which are deformed by the strain.

3.3 ADVANCED TEM CHARACTERIZATION OF PEROVSKITE MATERIALS: A STRAIGHTFORWARD METHOD TO ANALYZE NOVEL CATION ORDERED MATERIALS

Compounds with general structural formula ABX_3 commonly display a perovskite structure. These structures have received a great deal of scientific interest due to their interesting physical properties, such as superconductivity, ferroelectricity, colossal magnetoresistance, piezoelectricity, nonlinear optical behavior, and so forth. The perovskite basic structure consists of corner-sharing octahedral of X anions (O, F) with the B cation at their centers forming BO_6 octahedra. The A cation is situated in between vertices shared BO_6 octahedra. Oxygen-deficient perovskites ABO_{3-x} represent a huge source for the generation of new physical properties in view of future applications, going from materials for energy storage such as solid oxide fuel cell cathodes (SOFCs) to magnetic, multiferroic, magnetoresistive, or superconducting materials for various electronic devices. The possibility of tuning the oxygen stoichiometry is of primordial importance for the optimization of the performances of those oxides. Introducing oxygen vacancies leads not only to possible structural changes (tilting, displacement, and transformation of the anion octahedra) but, what is of primary importance, a change of oxidation state of the B cation. Importantly, the great flexibility of this framework allows the simultaneous introduction of two different cations with different valences and sizes, as well A sites and B sites, which leads to structural distortions, chemical ordering, and the presence of similar cations with different valences within the same structure. This can induce complex types of cationic and concomitantly anionic orderings at a nanoscale, significantly modifying the properties in this way. However, the concomitant structural changes often are so small that the splitting of characteristic reflections in the XRD pattern due to distortion or superstructure ordering reflections are hardly detectable; it is similar for local chemical (chemical ordering) and electronic (valences) structures. It is clear that an understanding of the structural, chemical, and electronic changes at the atomic level is of great importance, and advanced EM is the only method used to get all this information from single atomic columns simultaneously. Thus, the control and understanding of such properties require a systematic structural investigation of these oxides at a nanoscale, combining HRTEM, HAAD-STEM together with ABF-STEM, EDX elemental atomic mapping, and EELS. An example of such complex studies by means of advanced TEM that led to the discovery of a new family of perovskite material is presented in the following sections.

3.3.1 Quintuple Perovskites Revealed by Advanced TEM

3.3.1.1 Cation Ordering in $Sm_{2-\varepsilon}Ba_{3+\varepsilon}$ Fe_5 $O_{15-\delta}$ Complex Perovskite

It was demonstrated that ordering of cations in the form of layers in oxygen-deficient perovskites in $LnBaCo_2O_{5+\delta}$, the so-called "112" cobaltites (Raveau and Seikh 2012) and iron oxygen-deficient perovskites, such as $LnBaFe_3O_{8+\delta}$ and $LnBaFe_2O_{5+\delta}$ (Karen et al. 2002; Moritomo et al. 2003), is at the origin of attractive magnetic and magnetoresistance properties. In this respect, the oxygen-deficient perovskite $Sm_{2-\varepsilon}Ba_{3+\varepsilon}Fe_5O_{15-\delta}$ was synthesized and characterized by a number of techniques including advanced TEM. The structure and phase composition was controlled by XRD measurements, and the oxygen content in the single-phase oxide was determined by the thermogravimetric analysis (TGA) method (Volkova et al. 2014). Figure 3.11 summarizes the XRD and TEM results obtained for $Sm_{2-\varepsilon}Ba_{3+\varepsilon}Fe5O_{15-\delta}$.

The XRD pattern of $Sm_{2-\varepsilon}Ba_{3+\varepsilon}Fe5O_{15-\delta}$ shows characteristic peaks of cubic perovskite structure (Figure 3.11a) and can be unambiguously indexed based on the *Pm3m* space group, with a = 3.934(1) Å. The oxygen contents were found to be deficient with $\delta = 0.15 \pm 0.01$, which suggest a statistical random distribution of the Ba^{2+} and Sm^{3+} cations together with oxygen vacancies, with a cubic structure similar to what was reported previously in the literature for oxygen-deficient perovskites (Garcí-Gonzalez et al. 1993; Świerczek et al. 2011).

However, ED studies of this material (Figure 3.11b) show clear evidence of strong superstructure spots along one <100> cubic crystallographic direction implying increasing one of the unit cell parameter as $a_p = 5a_p$, corresponding to an "$a_p \times a_p \times 5a_p$" tetragonal cell. The HAADF-STEM image of the $Sm_{2-\varepsilon}Ba_{3+\varepsilon}Fe_5O_{15-\delta}$ structure along the [100] direction (Figure 3.11c) clearly confirmed cationic ordering in the (001) layers along the c-axis. One indeed observes rows of bright dots with three sorts of different intensities along the c-axis, as evident from the intensity scale profile in the Figure 3.11c insert. Bearing in mind the direct correspondence of the contrast and the atomic number Z, the possible structure models can be judged from their intensity, where high bright intensity layers correspond to Sm (Z = 62), less bright layers to Ba (Z = 56), and intermediate brightness layers are a mixture of Sm and Ba cations. Thus, based on the HAADF-STEM image the tetragonal structure can be directly interpreted by the following periodic stacking sequence of the A cationic layers along the *c*-axis: "Sm-Ba-Sm/Ba-Sm/Ba-Ba-Sm" caused the quintupling of the perovskite structure along one direction as it was observed in the ED pattern. Importantly, the presence of a dark line between two successive mixed Sm/Ba layers due to slightly different interplanar distances with respect to other layers ("Sm-Ba" or "Ba-Sm/Ba") is a signature of the present of oxygen vacancies.

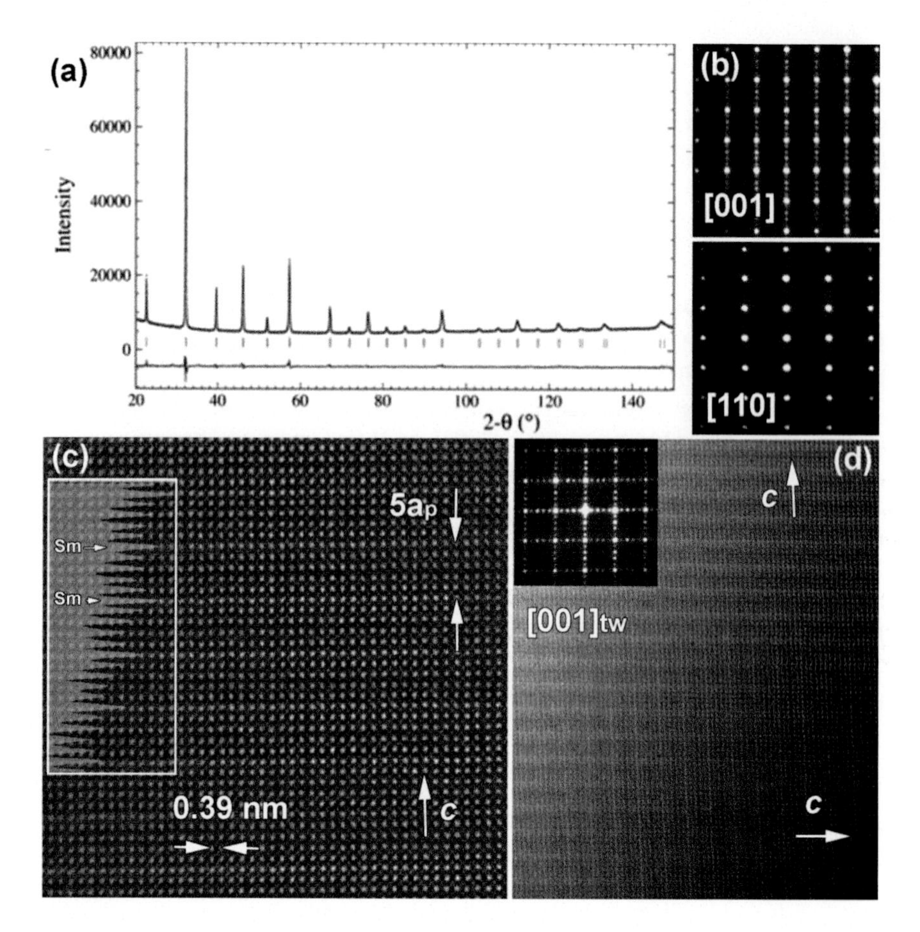

FIGURE 3.11 (a) Powder XRD pattern of $Sm_{2-\varepsilon}Ba_{3+\varepsilon}Fe_5O_{15-\delta}$ unambiguously indexed using the cubic *Pm3m* space group, with $a = 3.934(1)$ Å. (b) ED patterns of the $Sm_{2-\varepsilon}Ba_{3+\varepsilon}Fe_5O_{15-\delta}$ structure obtained along the most relevant [001], [110] zone axis orientations that look different from the cubic structure. (c) High-resolution [001] HAADF-STEM image of the $Sm_{2-\varepsilon}Ba_{3+\varepsilon}Fe_5O_{15-\delta}$ structure showing a $5a_p$ contrast periodicity along the *c*-axis. Line intensity scan profile along the *c*-direction of the HAADF-STEM image is overplayed on the image. The intensity of the peaks indicates an Sm-Ba-Ba/Sm-Ba/Sm-Ba-Sm ordering, where higher intensity peaks correspond to Sm, less intense to Sm/Ba mixture columns, and lowest intensity peaks are Ba columns. (d) High-resolution [001]HAADF-STEM image of a 90-degree twinned area with the corresponding ED pattern seen in the insert.

The HAADF-STEM images and corresponding ED pattern show that material is hardly chemically twinned (Figure 3.11d) along the three cube directions of the perovskite basic structure, which is a typical feature of tetragonal structures.

To confirm the proposed HAADF-STEM chemical ordering, elemental EELS was applied. EELS elemental mapping is illustrated in Figure 3.12 and demonstrates that the brighter layer consists only of Sm atoms, spaced by $5a_p$ along the c-axis. The neighborhood less intensity layers are pure Ba, and the intermediate intensity layers are intermixed Ba and Sm in approximately equal amounts of each element (Figure 3.12a). The Fe sublattice is intact by the ordering and undisturbed throughout the full superstructure. All these data confirm the structure proposed from HAADF-STEM intensity analysis.

The last question is the localization of oxygen position, possible oxygen vacancies, and corresponding changes of valence and bonding coordination. Because the fine structure of EELS edges is found to be sensitive to the local environment of elements (Tan et al. 2012; Turner et al. 2012), the spatially resolved electron energy-loss at high energy resolution was applied. The spatially resolved EELS data for O-K and Fe-$L_{2,3}$ edges are presented in Figure 3.12b and revealed that the O-K edge spectra corresponding to the "FeO_2" planes (labeled I, II, III) exhibit different intensity ratios of the two prepeaks (*a* and *b*) to the O-K edge, depending on the surrounding Sm/Ba layers. The O-K fine structure in the Sm layer is very similar to that of plane I, whereas those of the Ba and Ba/Sm layers are similar to planes II and III, correspondingly. Two prepeaks in O-K edge spectra are related to different states: prepeak *a* can be attributed to a charge transfer from the eg to the t2g band of Fe, and prepeak *b* is related to Fe3d eg-O2p hybridized states. Prepeak *b* is almost invariant in the structure, whereas the height of prepeak *a* varies from the layers. Increasing prepeak *a* related to Fe3d t2g-O2p hybridized states is clearly visible in the Ba/Sm mixed layers. These data supported the presence of oxygen vacancies in those planes and the fact that prepeak is stronger in the panel III layer, suggesting the presence of more vacancies in this plane, which is in good agreement with the HAADF-STEM observation.

The position of oxygen atoms was emphasized using high-resolution ABF-STEM imaging along two main crystallographic zone axes that are acquired simultaneously with HAASDF-STEM. The ABF-STEM images in Figure 3.13 reveal that the oxygen positions in all the layers are close to the ideal octahedral positions, and no vacancies have been observed. However, a careful analysis of the images reveals that the oxygen columns in the equatorial positions close to the Sm layer are slightly shifted from their ideal octahedral position to the direction of Sm^{3+} cations yielding a "zigzag" contrast along [100] and [110]. The absence of visible oxygen vacancies can be explained by random distribution of oxygen vacancies within the layer and projected nature of the images in TEM.

The discovery a new type of oxygen-deficient perovskite material by advanced TEM—quintuple $Sm_{2-\varepsilon}Ba_{3+\varepsilon}Fe_5O_{15-\delta}$ perovskite—stimulated the search for other possible members of quintuple perovskites with potentially different compositions. To understand the role of the effect of the size of the Ln^{3+} cation on the ordering of the layers and on the chemical twinning at a nanoscale, a number of new compounds with Ln = Eu, Nd, Pr, Y... oxides were synthesized. Also, the presence of

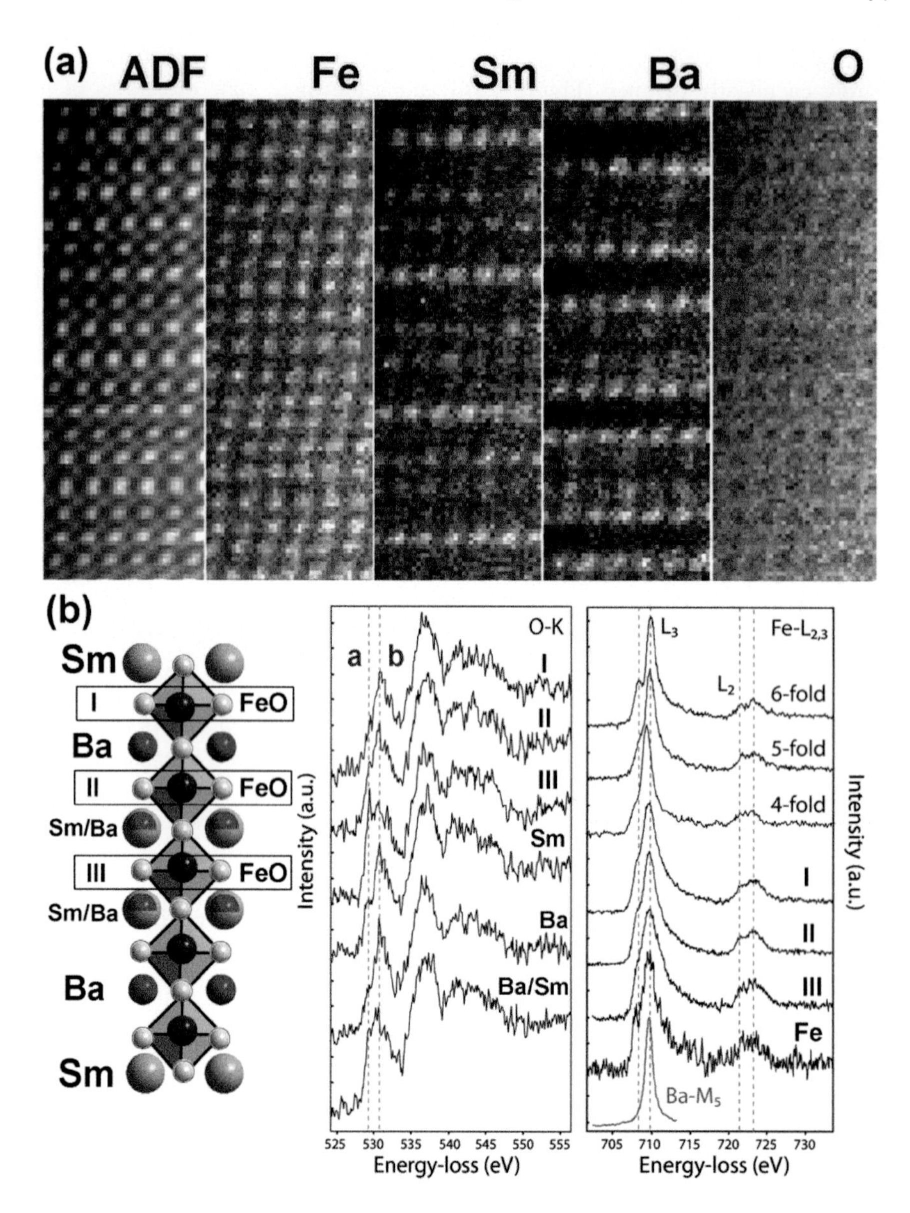

FIGURE 3.12 (a) Overview ADF-STEM image and EELS elemental mapping of $Sm_{2-\varepsilon}Ba_{3+\varepsilon}Fe_5O_{15-\delta}$ for Fe-$L_{2,3}$, sm-$m_{4,5}$ Ba-$M_{4,5}$, and O-K. (b) Corresponding structural model (left panel) together with the ELNES fine structure of $Sm_{2-\varepsilon}Ba_{3+\varepsilon}Fe_5O_{15-\delta}$. The O-K edge fine structure in the middle panel is a signature from the regions indicated in the left panel, and the Fe-$L_{2,3}$ fine structure in the right panel is a signature from the regions indicated in the left panel with references for fourfold, fivefold, and sixfold coordinated Fe^{3+} and a simultaneously acquired and energy-shifted Ba M_5 edge.

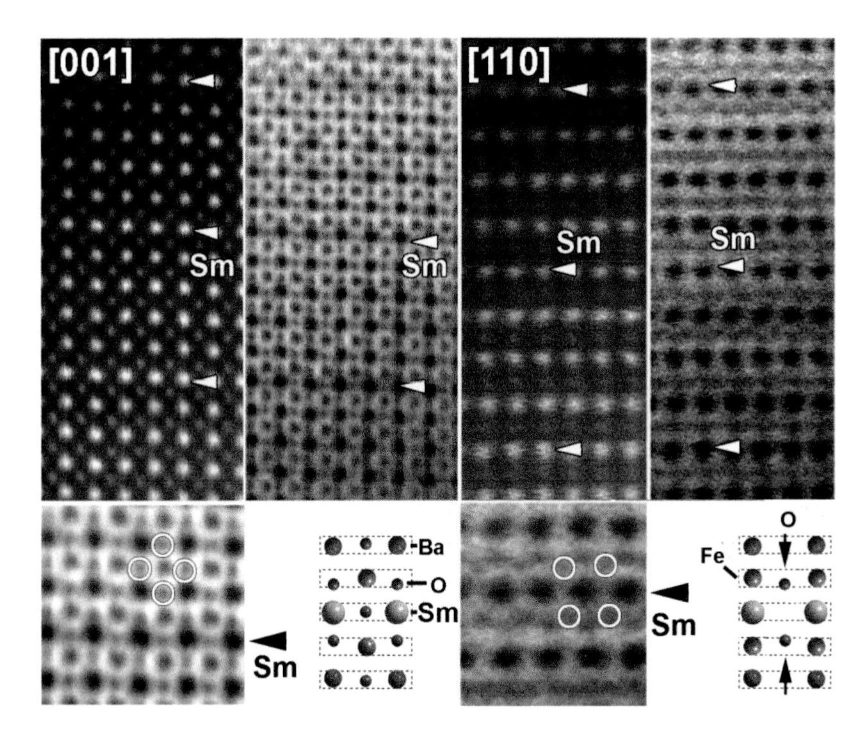

FIGURE 3.13 High-resolution [001] and [110] HAADF-STEM and simultaneously acquired ABF-STEM images of $Sm_{2-\varepsilon}Ba_{3+\varepsilon}Fe_5O_{15-\delta}$. High-magnification ABF-STEM images show the displacement of oxygen close to the Sm columns (marked with white circles). This displacement takes place with respect to the standard octahedral symmetry positions along the *c*-axis, as is shown in the corresponding structural model.

two different cations on the B sites, such as iron and cobalt, may also influence the structure and may produce other orderings and new properties of material.

3.3.1.2 Complex Ordering of Eu and Ba in $Eu_2Ba_3Fe_3Co_2O_{15-\delta}$ Quintuple Oxygen-Deficient Perovskite

Based on the above mentioned considerations, the $Eu_2Ba_3Fe_3Co_2O_{15-\delta}$ ($\delta \sim 1.28$) structure was investigated by advanced TEM (Kundu et al. 2015). It should be noted that XRD does not emphasize any deviation from the perovskite structure and has no superstructure presence. In contrast, the HAADF-STEM image of the $Eu_2Ba_3Fe_3Co_2O_{15-\delta}$ structure and corresponding ED pattern along the [100] direction presented in Figure 3.14 shows that $Eu_2Ba_3Fe_3Co_2O_{15-\delta}$ clearly looks like $Sm_{2-\varepsilon}Ba_{3+\varepsilon}Fe_5O_{15-\delta}$ quintuple perovskite structure: Eu and Ba cationic ordering in (001) layers along the *c*-axis. The combination of HAADF-STEM (Figure 3.14a) imaging and EELS elemental mapping presented in Figure 3.14d shows typical quintuple perovskite structure A-cation ordering (Figure 3.14c), which is characterized by the following layer stacking: Eu-Ba-Eu/Ba-Eu/Ba-Ba-Eu. No Fe/Co ordering is found suggesting random homogeneous distribution of iron and cobalt over the structure.

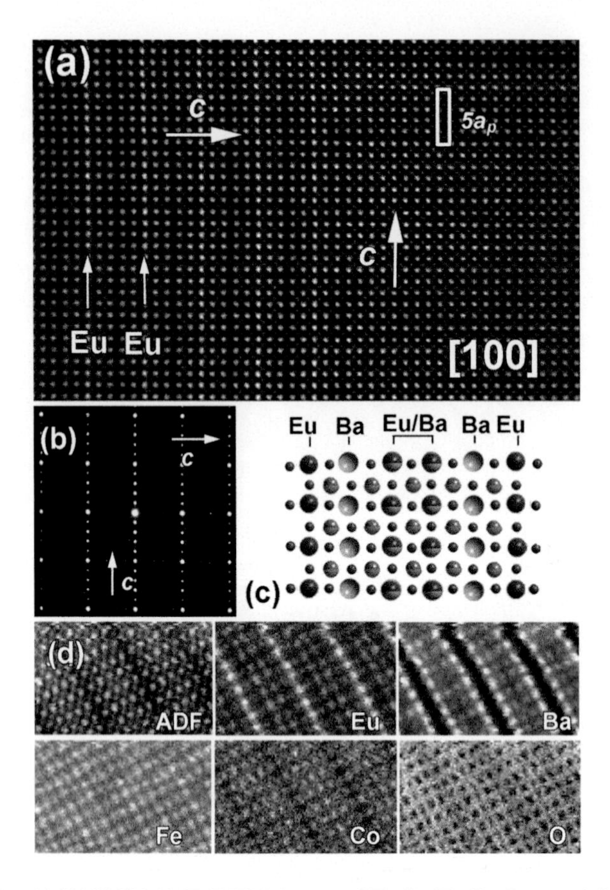

FIGURE 3.14 (a) [100] HAADF-STEM image of $Eu_2Ba_3Fe_3Co_2O_{15-\delta}$ and (b) corresponding ED pattern. (c) Structural model of $Eu_2Ba_3Fe_3Co_2O_{15-\delta}$ derived from the HAADF-STEM image in (a) and elemental the EELS mapping presented in (d).

The [100] ABF-STEM high-resolution image shown in Figure 3.15 gives more information about oxygen. Different from $Sm_{2-\varepsilon}Ba_{3+\varepsilon}Fe_5O_{15-\delta}$, the anionic sites within the Eu layers (labeled A in Figure 3.15a insert) are almost unoccupied (confirmed by the maximum at the O site on the C line intensity profile in the Figure 3.15a insert). Differently, the substantial amounts of oxygen vacancies appear within the $(Fe/Co)O_2$ layers sandwiched between one BaO and one mixed AO layer (labeled B in the Figure 3.15a insert), as shown from the shallow minimum at O site (A line intensity profile in the Figure 3.15 insert). In contrast, no oxygen vacancy is observed in the $(Fe/Co)O_2$ layers sandwiched between two mixed AO layers (labeled A in the Figure 3.15a insert), as shown from the deeper minimum at O site on the A line intensity profile. From Figure 3.15b it is immediately apparent that the oxygen positions between the Eu and Ba layers are strongly displaced toward the Eu columns along *c* with respect to the standard octahedral positions yielding a zigzag contrast.

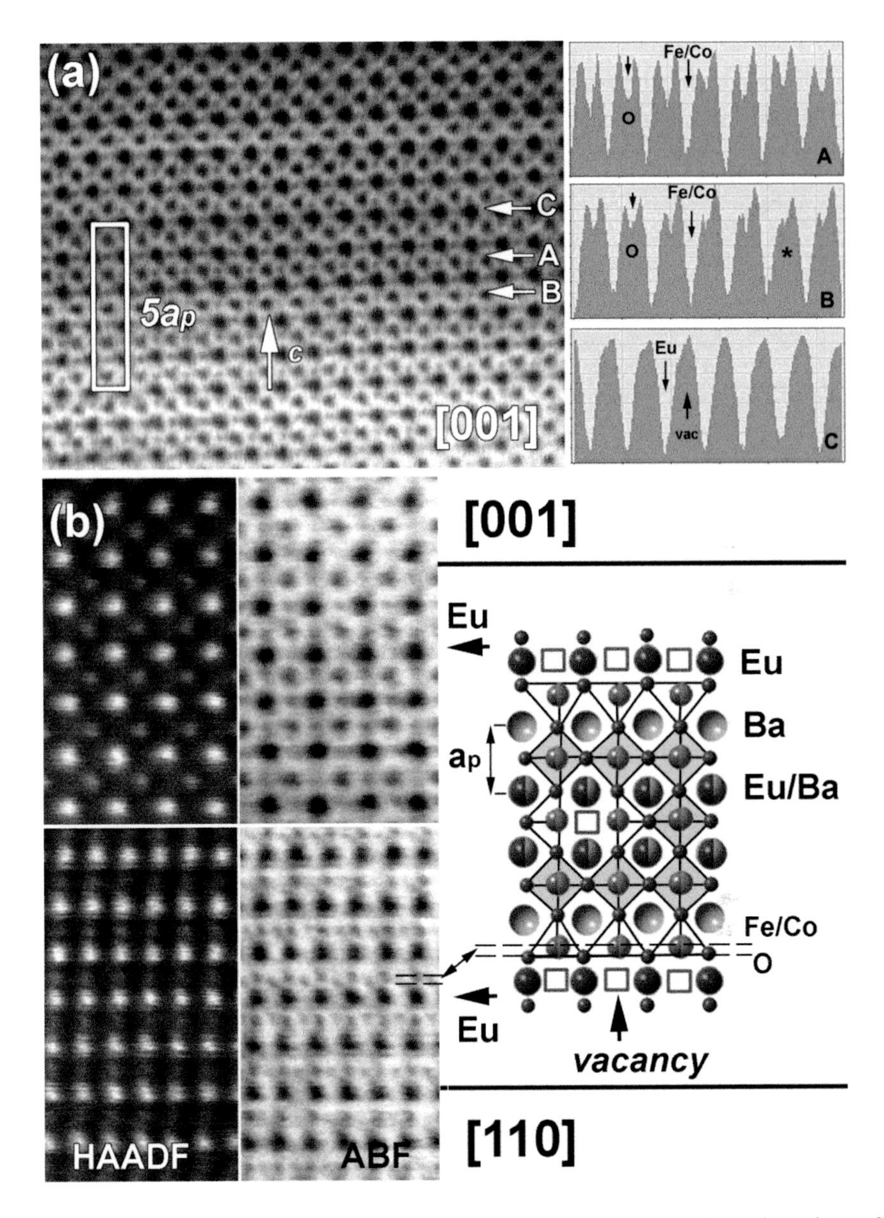

FIGURE 3.15 (a) ABF-STEM images along the [001] zone axis orientation of $Eu_2Ba_3Fe_3Co_2O_{13.72}$ showing the oxygen lattice. Line intensity profile over the A direction along the Eu layer (C), along next to the BaO Fe/Co-O layer (a) and along the Fe/Co-O layer in between (Ba/Eu) mixed layers are in the right panel. The deep minimum corresponds to the position of Eu and Fe/Co atomic columns (dark black contrast dots), whereas the local minimum indicates O position (fine gray dots). (b) Enlarged HAADF-STEM and simultaneously acquired ABF-STEM images of $Eu_2Ba_3Fe_3Co_2O_{13.72}$ along the [001] and [110] zone axis and corresponding structural model. The oxygen positions close to the Eu columns are severely displaced with respect to the standard octahedral symmetry positions along the *c*-axis. Oxygen vacancies are evident from the [001] ABF-STEM image.

The complete structure of $Eu_2Ba_3Fe_3Co_2O_{13.72}$ derived from HAADF-STEM and ABF-STEM studies is represented in Figure 3.15 in the right panel.

The elemental EELS mapping is represented in Figure 3.16 and confirms the quintuple structure of $Eu_2Ba_3Fe_3Co_2O_{15-\delta}$ perovskite. The Ba and the Eu maps (Figure 3.16b and c, correspondingly) show that the (001)-Eu layers (red dot rows) are sandwiched between two Ba layers (green bright dots) and are separated by $5a_p$. The contrast between two successive Ba layers (Figure 3.16a) (spaced by $3a_p$) confirms that they are separated by two mixed A (Ba/Eu) layers (green + dark spots). However, the Ba and Eu cations are not distributed at random within those layers. One indeed observes on the colored Ba-Eu overlay figure (Figure 3.16a and f) that double red rows parallel to [010] corresponding to Eu (marked by white arrows) alternate with double green [010] rows corresponding to barium along *a*. Thus, these observations confirm that the stacking sequence of the Ba and Eu layers in the $Eu_2Ba_3Fe_3Co_2O_{15-\delta}$ quintuple perovskite is “Eu-Ba-A-A-Ba-Eu,” where the mixed layers (A) contain ~50% Ba and ~50% Eu, as previously described for Ln = Sm. Also, a double perovskite region is detected in the EELS elemental mapping (Figure 3.16a–c marked by white arrowheads), which is intergrowth within the quintuple perovskite structure.

Moreover, Ba and Eu are not distributed at random in the mixed layers, and the ordering of Eu and Ba is correlated in two successive mixed layers. Double “Ba_2” and “Eu_2” rows parallel to [010] are formed between two adjacent mixed layers; moreover, one Ba_2 double row alternates with one double Eu_2 row along the *a*-direction. Thus, the stacking sequence of the Ba and Eu layers in the Eu phase can be described as “Eu-Ba-$[Eu_2\text{-}Ba_2]_{0.5}$-Ba-Eu” (Figure 3.16f, insert). It is remarkable that this particular ordering of the mixed layers is not observed in the quintuple perovskite $Sm_2Ba_3Fe_5O_{15-\delta}$ containing only iron in the B sites, because no ordering of the Co and Fe cations can be detected. It is of interest to try to understand the relationships between the anionic vacancy and cationic ordering in those oxides. The EELS images of oxygen mapping (Figure 3.16e) show that the oxygen vacancies in this structure are not distributed at random but form double layers (darker double rows) spaced by $5a_p$ and located at the level of the mixed $[Eu_2\text{-}Ba_2]_{0.5}$ cationic layers. Finally, the chemical analysis of this compound would imply the mixed valence Fe^{3+}/Fe^{4+} and the presence of Co^{3+} according to the formula $Eu_2Ba_3Fe^{III}{}_{2.57}Fe^{IV}{}_{0.43}Co^{III}{}_2O_{13.72}$. The spatially resolved EELS spectra of this oxide at the FeL_3 and CoL_3 edge are shown in Figure 3.17. They confirm the presence of Co^{3+} and Fe^{3+} and show by the presence of a prepeak at ~707 eV that the charge distribution around iron does not correspond to Fe^{4+} but rather to Fe $3d^5$ L.

In conclusion, an advanced TEM study shows that $Eu_2Ba_3Fe_3Co_2O_{13.72}$ exhibits like the two other members of the series, Ln = Nd (Kundu et al. 2015a), Sm (Volkova et al. 2014), and a fivefold stacking of single Eu and Ba (001) layers with mixed “$Eu_{0.5}Ba_{0.5}$” layers. However, remarkably it differs from the latter because europium and barium are ordered within and between two successive mixed layers, forming double Ba_2 and Eu_2 columns running along [010], so that the distribution of Eu and Ba on the A sites can be de described as a combination of a layered and columnar ordering according to the stacking sequence Eu-Ba-$[Eu_2\text{-}Ba_2]_{0.5}$-Ba-Eu.

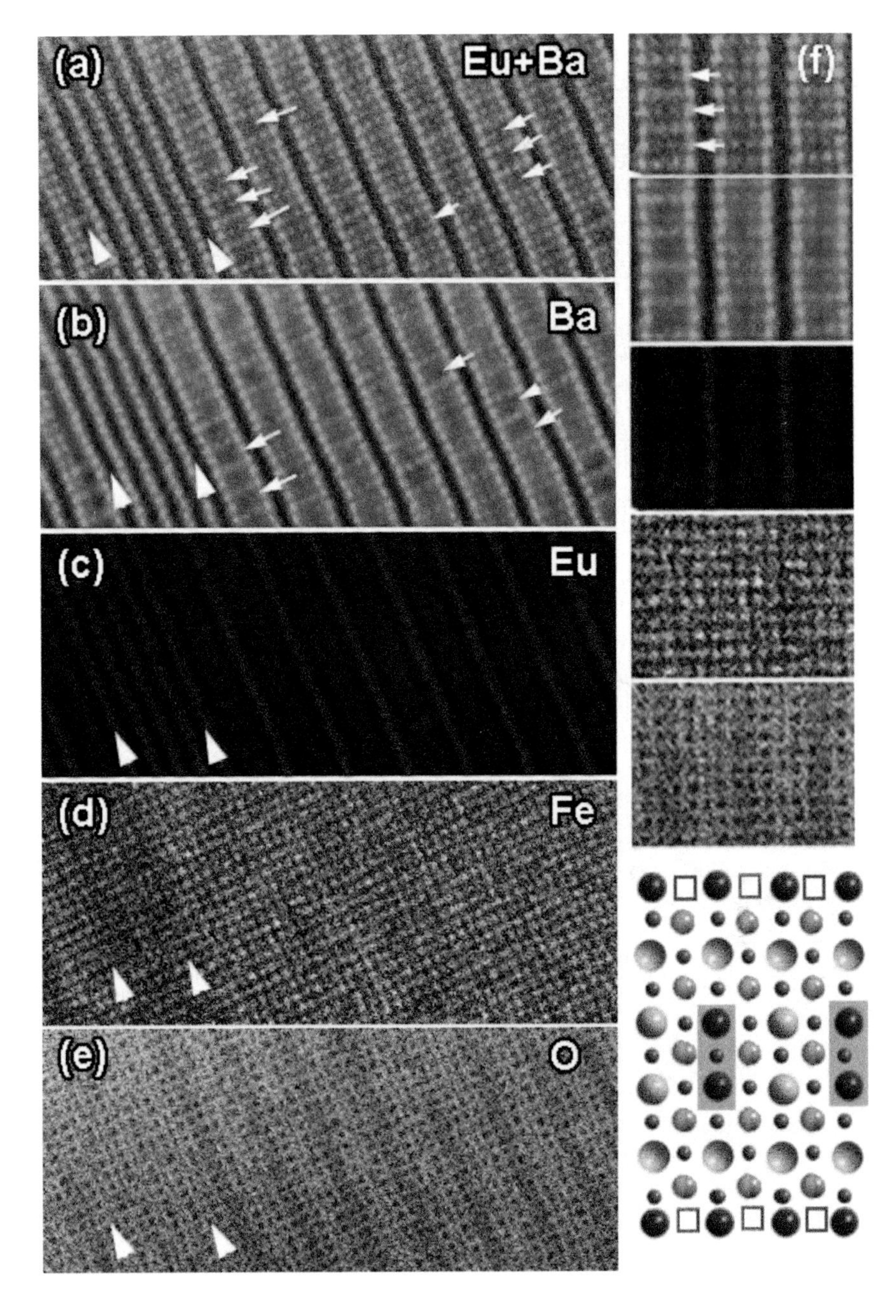

FIGURE 3.16 EELS elemental mapping of $Eu_2Ba_3Fe_3Co_2O_{13.72}$ quintuple perovskite. (a) Color overlay with Eu and Ba. Note the area in which Ba and Eu are ordered within mixed Eu/Ba layers marked by white arrows. (b) Ba map in green, (c) Eu-map in red, (d) Fe-$m_{4,5}$ map in yellow, and (e) O-K map in blue. A defect 112 region is often observed, depicted by arrowheads. (f) Enlargement of the mapping image of the Ba-Eu ordered part and corresponding structural model (turned by 90°), with Eu in purple, Ba in yellow, Fe in gray, Co in green, and O in blue.

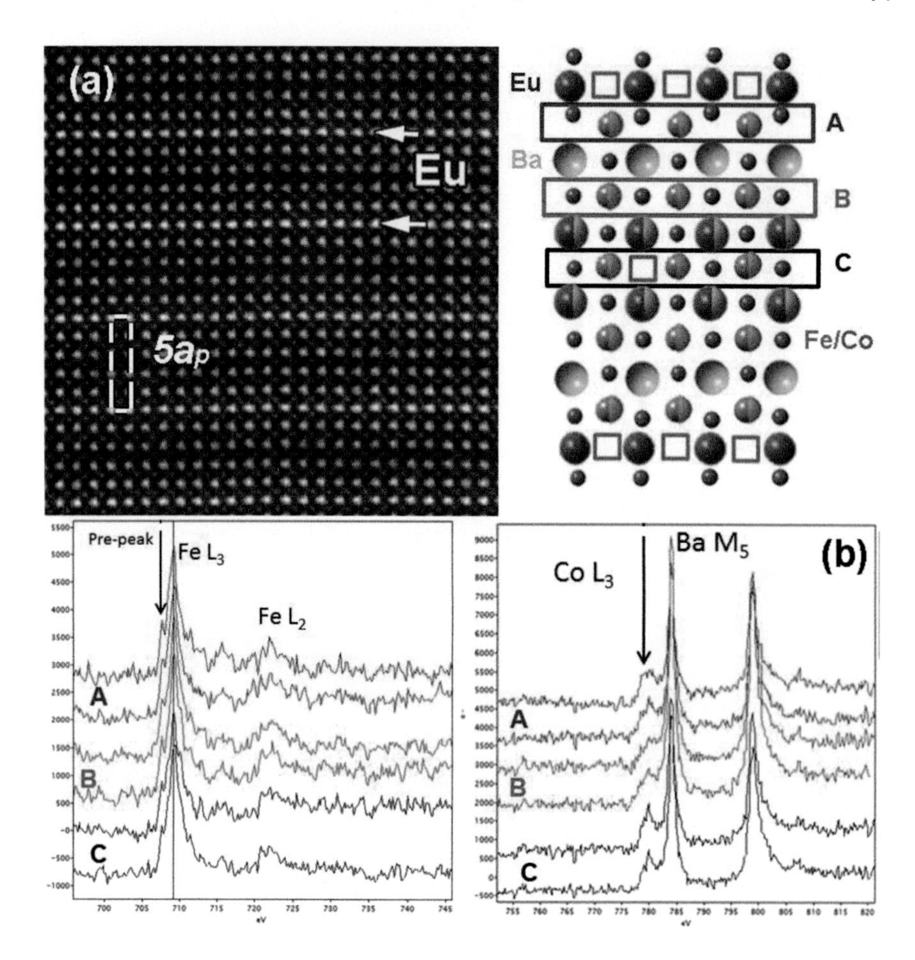

FIGURE 3.17 (a) HAADF-STEM image and structural model of $Eu_2Ba_3Fe_3Co_2O_{13.72}$ indicated particular layers of EELS-STEM scanning. (b) ELNES fine structure of $Sm_{2-\varepsilon}Ba_{3+\varepsilon}Fe_5O_{15-\delta}$. The Fe-$L_{2,3}$ fine structure is in the left panel signatures from the regions indicated in (a), A, B, and C, and a simultaneously acquired and energy-shifted Ba M_5 edge.

This unique cationic ordering appears to be closely related to the size of the lanthanide and correlatively to the anionic vacancy concentration. One indeed observes that the tendency to ordering increases as the Ln^{3+} size and oxygen content decreases. For the larger cations Nd and Sm, the mixed layers do not show any signature of ordering between barium and lanthanide, whereas the decrease of the size to Eu leads to this unusual columnar ordering between two mixed layers. If this conclusion is true, then it cannot exclude the possibility of complete ordering of the Ln and Ba layers in the quintuple oxygen-deficient perovskites for the smallest rare-earth content according to the sequence "Ba-Ln-Ba-Ba-Ln-Ba." Another question arises because Eu-Ba ordering is accompanied by oxygen vacancy ordering. Is it possible that A-cation ordering will stimulate B-cation ordering (Fe/Co in our case) through oxygen vacancy ordering?

3.3.2 Exceptional Layered Ordering of Cobalt and Iron in $Y_2Ba_3Fe_3Co_2O_{13.36}$ Oxygen-Deficient Perovskite Revealed by Advanced TEM

It should be noted that the idea of possible Fe/Co ordering in the perovskites seems quite unrealistic because the B-cation ordering in ABO_3 perovskite structure is a rare phenomenon. The second point is that Fe and Co are very similar in their ionic size and their ability to accommodate the same type of coordination. To get possible B-cation ordering, B cations should satisfy certain requests, such as different ionic size and valence. In most of the cases, they exhibit random distribution within the perovskite structure.

Layered perovskites Ln-Ba-Co-O and Ln-Ba-Fe-O have received a great deal of attention due to their magnetic and magnetoresistance properties (Raveau and Seikh 2012) and as a possible candidate for oxygen storage in SOFCs (Kim and Manthiram 2008; Świerczek 2011; Lekse et al. 2014). They are well studied and reported in the literature mainly to be the double perovskites for Co ($LnBaCo_2O_{5+\delta}$) (Maignan et al. 1999; Rautama et al. 2008) structure and double and triple perovskites in the case of Fe ($LnBaFe_2O_{5+\delta}$ and $LnBa_2Fe_3O_{8+\delta}$, respectively) (Garcí-Gonzalez et al. 1993; Karen and Woodward 1999). They consist of regular ordered layers of rare-earth (or lanthanide Ln) and barium cations. Moreover, the oxygen-deficient perovskites exhibit a specific ordering of their oxygen vacancies on the anionic sites as well.

Taking all this into consideration, the $Y_2Ba_3Fe_3Co_2$ compound was prepared in air and characterized by conventional structure characterization techniques. The powder XRD pattern shows that similar to all quintuple perovskites, the XRD pattern exhibits peaks that can be indexed in a cubic perovskite with the *Pm3m* space group and $a = a_p$ (Lebedev et al. 2016). Similar to all quintuple oxygen-deficient perovskites, ED patterns along the main cubic perovskite zone axis show the presence of fivefold superstructure reflections only along one a_p direction with respect to the basic cubic subcell, corresponding to the tetragonal perovskite structure with $a = b = a_p$, $c = 5a_p$ cell parameters, as seen in Figure.3.18a. However, the careful analysis of ED patterns revealed that the superstructure reflections along the *c*-direction show streaks, and they are not exactly commensurate with a_p, indicating more complicated, rather than similar to quintuple perovskite, short-range ordering.

The [100] high-resolution HAADF-STEM and simultaneously acquired ABF-STEM images in Figure 3.18b clearly show $5a_p$ periodicity along the *c*-axis. However, the contrast caused by stacking of the Ba (Z = 56) and Y (Z = 39) layers is obviously different from the other Ln-quintuple perovskites (Volkova et al. 2014; Kundu et al. 2015a,b). One observes only two sorts of layers (very bright dots and much less bright dots corresponding to Ba and Y, respectively) with the stacking sequence "Y-Ba-Ba-Y-Ba-Y" along the *c*-axis, which is different from the other Ln-quintuple perovskites, which exhibit three sorts of layers, Ln, Ba, and mixed A ($Ba_{0.5}Ln_{0.5}$), with the stacking sequence "Ln-Ba-A-A-Ba-Ln."

Taking into account the obtained HAADF-STEM data, such stacking of $Y_2Ba_3Fe_3Co_2O_{13+\delta}$ can be described as an intergrowth of triple (Y-Ba-Ba-Y) and double (Y-Ba-Y) perovskites. The ABF-STEM imaging completed the structure of $Y_2Ba_3Fe_3Co_2O_{13+\delta}$ with position of oxygen and possible oxygen vacancies. One can see

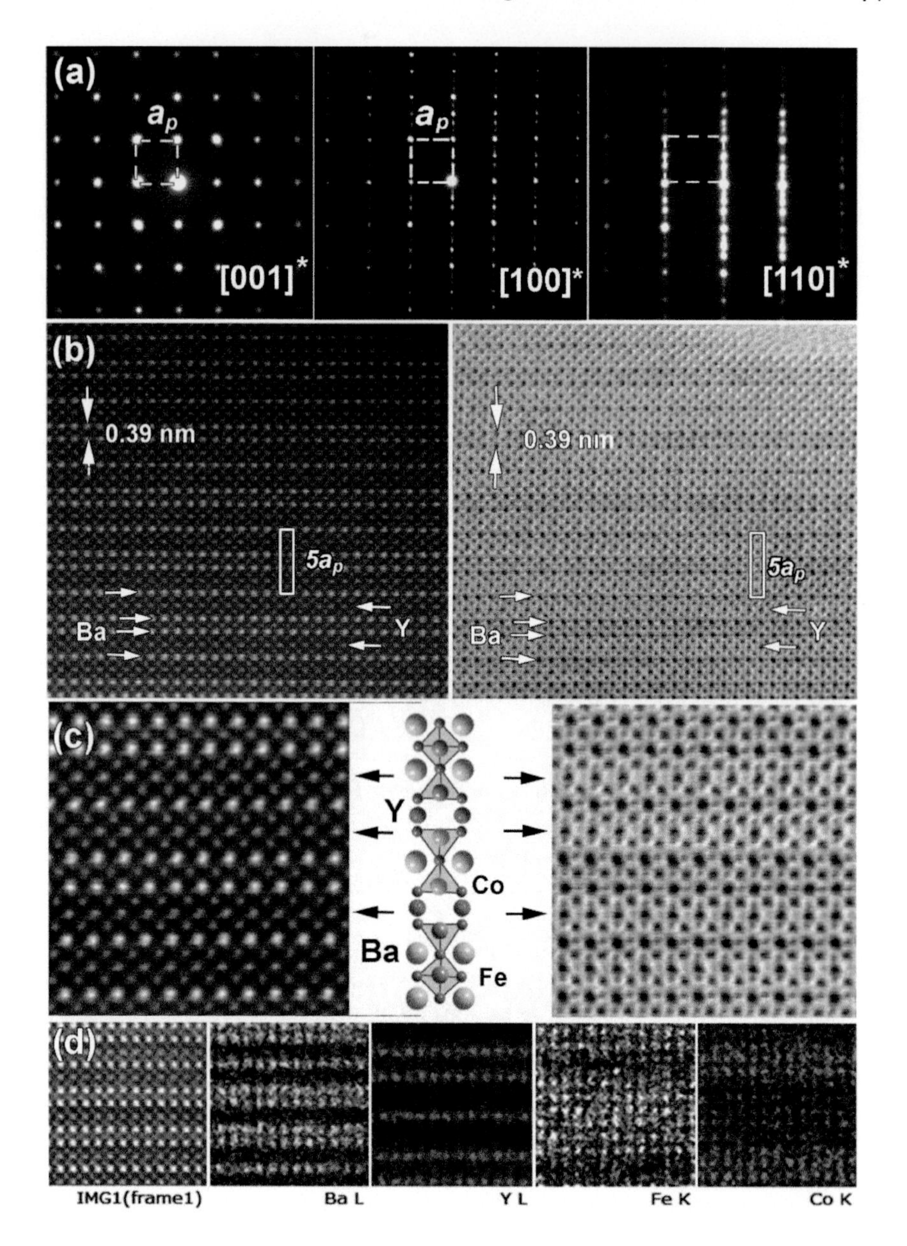

FIGURE 3.18 (a) ED patterns along the main zone axis of the $Y_2Ba_3Fe_3Co_2O_{13.36}$ structure: [001]*, [100]*, and [110]*. Note the presence of modulation along the *c*-axis in [100]* and [110]* ED patterns. (b) The [100] HAADF-STEM image and simultaneously acquired ABF-STEM high-resolution images of perfect ordered $Y_2Ba_3Fe_3Co_2O_{13.5}$. (c) Enlargement of HAADF-STEM (left panel) and ABF-STEM (right panel) images and corresponding structural model of $Y_2Ba_3Fe_3Co_2O_{13.5}$ (center panel). Note that oxygen vacancies are located between the Y columns along the *a*-axis and that the oxygen positions close to the Y columns are severely displaced along the *c*-axis. (d) HAADF-STEM image and atomic EDX elemental mapping of Ba L-, Y L-, Fe K-, and Co K-evidenced Co-Fe ordering.

from the high-magnification ABF-STEM image in Figure 3.18c that the anionic sites between two barium cations in Ba layers are fully occupied by oxygen, forming stoichiometric (001) BaO layers. While Y layers definitely exhibit oxygen vacancies, the anionic sites between two yttrium cations are very weakly occupied (weak and diffuse gray spots) or almost empty. Bearing in mind the data of iodometric titration and TGA (Lebedev et al. 2016), it can be concluded that oxygen vacancies are localized within the Y layer forming $YO_{\delta/2}$ layers with partially occupied anionic sites and $\delta \sim 0.36$.

The last question is what happens with Fe and Co in this structure? Due to the close atomic number of Fe (Z = 26) and Co (Z = 27), it is not possible to distinguished these two cations by HAADF-STEM imaging (Van Aert et al. 2009). Therefore, EELS elemental mapping was acquired to find simultaneously chemical and electronic information on $Y_2Ba_3Fe_3Co_2O_{13+\delta}$. Figure 3.19 displayed EELS results obtained from $Y_2Ba_3Fe_3Co_2O_{13+\delta}$.

The EELS elemental mapping confirms the perfect Ba-Y ordering suggested by HAADF-STEM imaging; double Ba layers (Ba2) are separated with single Ba layers (Ba1) along the *c*-axis by single Y layers (Figure 3.19, upper panel, an overplayed Ba&Y color image) . No intermixing of Y and Ba is detected. The most remarkable result from EELS mapping is evidence of the corresponding ordering of Fe and Co, which was not detected by HAADF-STEM. A perfect stacking of triple Fe layers (Fe3) with double Co layers (Co2) within a perovskite matrix is evident from EELS elemental mapping (Figure 3.19, upper panel, an overplayed Fe&Co color image). Such Fe/Co ordering within the perovskite structure was never observed before and looks very unusual in spite of the similar size of Co and Fe and their ability to accommodate the same type of coordination. The atomic EELS elemental mapping unambiguously shows that the double Ba2 layers are separated by the triple Fe3 layers, whereas the single Ba1 layers are sandwiched between the double Co2 layers, meaning Fe prefers triple perovskite and Co prefers double perovskite structures. Finally, this B site cationic ordering can be described as following the stacking sequence along the *c*-axis: "Fe-Ba-Fe-Ba- Fe-Y-Co-Ba-Co-Y-Fe." This exceptional type of cationic ordering suggests that the oxygen content and oxygen vacancy distribution play a role in its stabilization, resulting from the alternate stacking of the double $YBaCo_2O_5$ perovskite cell and the $YBa_2Fe_3O_8$ triple perovskite cell along the *c*-axis. The ideal composition of such a triply ordered perovskite will be $Y_2Ba_3Fe_3Co_2O_{13}$. This is close to that observed experimentally, which is $Y_2Ba_3Fe_3Co_2O_{13+\delta}$, with $\delta \sim 0.36$.

To understand the charge distribution in this ordered perovskite, spatially resolved EELS have been performed on perfectly ordered crystals (Figure 3.19, bottom panel) along three different Fe layers and along the Co layer, as indicated in the insert of selected HAADF-STEM images from Figure 3.19 (upper panel, white boxes)

From the absolute position of the Fe-$L_{2,3}$ and Co-$L_{2,3}$ edges, it is clear that both Fe and Co are in the trivalent state (Tan et al. 2012; Turner et al. 2012). The absence of a clear prepeak to the Fe-L_3 edge indicates a deviation from a perfect octahedral coordination toward a lower coordination for iron (Turner et al. 2012). This is most likely caused by the presence of oxygen vacancies. All these data strongly suggest that the founded ordering of Co and Fe in the form of layers is strongly dependent on the oxygen stoichiometry.

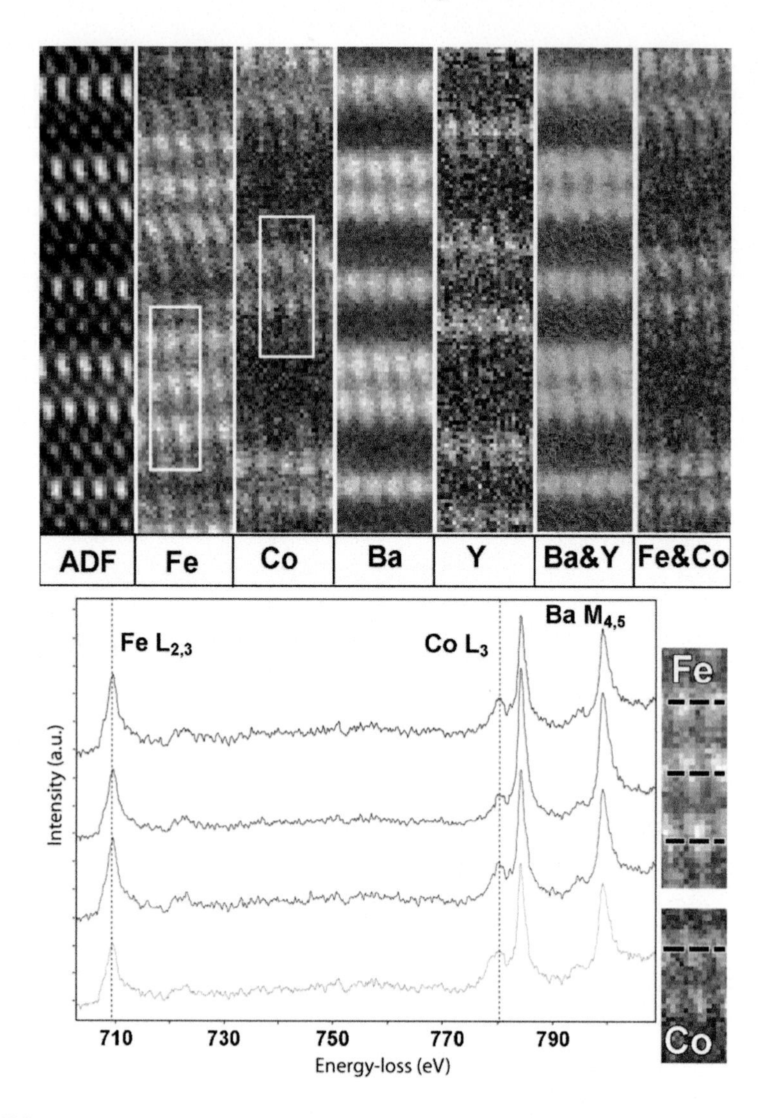

FIGURE 3.19 Top: HAADF-STEM image and EELS elemental mapping of ordered $Y_2Ba_3Fe_3Co_2O_{13.5}$ for Fe-$L_{2,3}$, Co-$L_{2,3}$, Ba-$M_{4,5}$, and O-K. Color overlays Ba & Y (Y in red and Ba in green) and Fe & Co (Fe in red and Co in green). Bottom: Corresponding EELS fine structure obtained from the regions indicated as Fe and Co in the mapping.

Finally, by means of advanced TEM a new type of structure was discovered that corresponds to the perfect intergrowth [$YBa_2Fe_3O_8$] [$YBaCo_2O_5$] of the 123 ferrite and the 112 cobaltite, with oxygen excess (δ) distributed at the boundary of the two blocks, sandwiched between FeO_5 and CoO_5 pyramids.

REFERENCES

Abraham, F. and Mentre, O. 1994. Bi1. 7V8O16: the first bi-hollandite-type compound. *Journal of Solid State Chemistry, 109*(1): 127–133.

Ahn, C.C., 2004. *Transmission Electron Energy Loss Spectrometry in Material Science and EELS Atlas*. Berlin: Wiley-VSH Verlag GmbH.

Amelinckx, S., Van Dyck, D., Van Landuyt, J., and Tendeloo, G.V., 1997. *Handbook of Microscopy*. Berlin: Wiley-VSH Verlag GmbH

Banhart, F., 2008. *In-Situ Electron Microscopy at High Resolution*. Singapore: World Scientific Publishing.

Barrier, N., Lebedev, O.I., Seikh, M.M., Porcer, F., and Raveau, B., 2013. Impact of Mn^{3+} upon structure and magnetism of the perovskite derivative $Pb_{2-x}Ba_xFeMnO_5$ (x ~ 0.7). *Inorganic Chemistry, 52*: 6073–6082.

Beck, V.D., 1979. Hexapol spherical-aberration corrector. *Optik, 53*: 241–255.

Bertaut, F., Blum, P. and Sagnieres, A., 1959. Structure du ferrite bicalcique et de la brownmillerite. *Acta Crystallographica, 12*: 149–59.

Borisevich, A.Y., Lupini, A.R. and Pennycook, S.J., 2006. Depth sectioning with the aberration-corrected scanning transmission electron microscope. *Proceedings of the National Academy of Sciences, 103*(9): 3044–3048.

Carter, C.B., and Williams, D.B., 2016. *Transmission Electron Microscopy: Diffraction, Imaging and Spectrometry*. Dordrecht: Springer.

Champnes, P.E., 2001. *Electron Diffraction in the Transmission Electron Microscope*. United Kingdom: Taylor & Francis Ltd .

Chandler, J.A., 1977. *X-Ray Analysis in the Electron Microscope*. Amsterdam: Elsevier/ North-Holland Biomedical Press.

Chapman, J.N., 1984. The investigation of magnetic domain structures in thin foils by electron microscopy. *Journal of Physics D: Applied Physics, 17*(4): 623.

Chuvilin, A. and Kaiser, U., 2005. On the peculiarities of CBED pattern formation revealed by multislice simulation. *Ultramicroscopy, 104*(1): 73–82.

Cowley, J.M., 1992a. *Electron Diffraction Techniques*. Oxford : Oxford Univ. Press.

Cowley, J.M., 1992b. Twenty forms of electron holography. *Ultramicroscopy, 41*(4): 335–348.

De Graef, M., 2002. *Introduction to ConventionalTransmission ElectronMicroscopy*. Cambridge: Cambridge Univ. Press.

Dyck, O., Kim, S., Kalinin, S.V. and Jesse, S., 2017. Placing single atoms in graphene with a scanning transmission electron microscope. *APL, 111*: 113104.

Egerton, R.F., 1996. *Electron Energy-Loss Spectroscopy in the TEM*, New York: Plenum Press.

Egerton, R.F., 2009. *Electron Energy-Loss Spectroscopy in the TEM*, New York: Plenum Press. *Reports on Progress in Physics*: 016502.

Erni, R., 2010. *Aberration-Corrected Imaging in Transmission Electron Microscopy*, London: Imperial College Press.

Erni, R., Rossell, M.D., Kisielowski, C. and Dahmen, U., 2009. Atomic-resolution imaging with a sub-50-pm electron probe. *Physical Review Letters, 102*(9): 096101.

Findlay, S.D., Shibata, N., Sawada, H., Okunishi, E., Kondo, Y., Yamamoto, T. and Ikuhara, Y., 2009. Robust atomic resolution imaging of light elements using scanning transmission electron microscopy. *Applied Physics Letters, 95*(19): 191913.

Flannigan, D.J., *and* Zewail, A.H., 2012. 4D electron microscopy: principles and applications. *Accounts of Chemical Research, 45*(10): 1828–1839.

Frank, J., 1992. *Electron Tomography: Three-Dimensional Imaging with the Transmission Electron Microscope*, New York: Plenum Press.

Friedrich, H., de Jongh, P.E., Verkleij, A.J. and de Jong, K.P., 2009. Electron tomography for heterogeneous catalysts and related nanostructured materials. *Chemical Reviews, 109*(5): 1613–1629.

Fuller, H.W. and Hale, M.E., 1960. Domains in thin magnetic films observed by electron microscopy. *Journal of Applied Physics, 31*(10): 1699–1705.

Fultz, B., and Howe, J., 2013. *Transmission Electron Microscopy and Diffractometry of Materials*, Dordrecht: Springer.

Garcí-González, E., Parras, M., Gonzlez-Calbet, J.M. and Vallet-Regí, M., 1993. A new "123" family: $LnBa_2Fe_3O_z$: III. Ln = Dy, Ho. *Journal of Solid State Chemistry, 104*(2): 232–238.

Goodman, P.T. and Lehmpfuhl, G., 1968. Observation of the breakdown of Friedel's law in electron diffraction and symmetry determination from zero-layer interactions. *Acta Crystallographica Section A: Crystal Physics, Diffraction, Theoretical and General Crystallography, 24*(3): 339–347.

Greaves, C., Jacobson, A.J., Tofield, B.C. and Fender, B.E.F., 1975. A powder neutron diffraction investigation of the nuclear and magnetic structure of $Sr_2Fe_2O_5$. *Acta Crystallographica Section B: Structural Crystallography and Crystal Chemistry, 31*(3): 641–646.

Guo, Q., Kim, S.J., Kar, M., Shafarman, W.N., Birkmire, R.W., Stach, E.A., Agrawal, R. and Hillhouse, H.W., 2008. Development of CuInSe2 nanocrystal and nanoring inks for low-cost solar cells. *Nano Letters, 8*(9): 2982–2987.

Haider, M., Rose, H., Uhlemann, S., Schwan, E., Kabius, B. and Urban, K., 1998. A spherical-aberration-corrected 200 kV transmission electron microscope. *Ultramicroscopy, 75*(1): 53–60.

Haine, M.E. and Mulvey, T., 1952. The formation of the diffraction image with electrons in the gabor diffraction microscope. *JOSA, 42*(10): 763–773.

Halliday, J.S. and Newman, R.C., 1960. Reflexion electron microscopy using diffracted electrons. *British Journal of Applied Physics, 11*(4): 158.

Hÿtch, M. and Houdellier, F., 2006. Mapping stress and strain in nanostructures by high-resolution transmission electron microscopy. *Microelectronic Engineering, 84/3*: 460–463.

Hÿtch, M., Snoeck, E. and Kilaas, R., 1998. Quantitative measurement of displacement and strain fields from HREM micrographs. *Ultramicroscopy, 74*: 131–146.

Jeong, S., Lee, B.S., Ahn, S., Yoon, K., Seo, Y.H., Choi, Y. and Ryu, B.H., 2012. An 8.2% efficient solution-processed CuInSe 2 solar cell based on multiphase CuInSe 2 nanoparticles. *Energy & Environmental Science, 5*(6): 7539–7542.

Kamada, R., Yagioka, T., Adachi, S., Handa, A., Tai, K.F., Kato, T., and Sugimoto, H., 2016, June. New world record Cu (In, Ga)(Se, S) 2 thin film solar cell efficiency beyond 22%. Paper presented at the 43rd IEEE Photovoltaic Specialists Conference (PVSC), Portland.

Karen, P. and Woodward, P.M., 1999. Synthesis and structural investigations of the double perovskites $REBaFe_2O_5$+w (RE = Nd, Sm). *Journal of Materials Chemistry, 9*(3): 789–797.

Karen, P., Woodward, P.M., Santhosh, P.N., Vogt, T., Stephens, P.W. and Pagola, S., 2002. Verwey transition under oxygen loading in $RBaFe_2O_5$+w (R = Nd and Sm). *Journal of Solid State Chemistry, 167*(2): 480–493.

Kim, J.-H. and Manthiram, A., 2008. $LnBaCo_2O_5$ oxide as cathodes for intermediate temperature solid oxide fuel cells. *Journal of the Electrochemical Society, 155*: B385.

Kirkland, E., 2010. *Advanced Computing in Electron Microscopy*. 2nd ed., Dordrecht: Springer.

Kisielowski, C., Erni, R., and Freitag, B., 2008. Object-defined Resolution below 0.5 Å in transmission electron microscopy-recent advances on the TEAM 0.5 instrument. *Microscopy and Microanalysis, 14*(S2): 78–79.

Koch, C.T., 2002. QSTEM package. Determination of Core Structure Periodicity and Point Defect along Dislocations. Ph.D. diss., Arizona State Univ.

Kossel, W. and Möllenstedt, G., 1939. Elektroneninterferenzen im konvergenten bündel. *Annalen der Physik*, *428*(2): 113–140.

Krakowiak, J., Lundberg, D. and Persson, I., 2012. A coordination chemistry study of hydrated and solvated cationic vanadium ions in oxidation states+ III,+ IV, and+ v in solution and solid state. *Inorganic Chemistry*, *51*(18): 9598–9609.

Krivanek, O.L., Dellby, N. and Lupini., A.R., 1999. Towards sub-angstrom electron beams. *Ultramicroscopy*, *78*: 1–11.

Krivanek, O.L., Lovejoy, T.C., Dellby, N. et al. 2014. Vibrational spectroscopy in the electron microscope. *Nature (London)*, *514*: 209.

Kundu, A., Mychinko, M., Caignaert, V., et al. 2015a. Quintuple perovskite phasoids in the system $Nd_{2-\varepsilon}Ba_{3+\varepsilon}(Fe,Co)_5O_{15-\delta}$ *JSSC*, *231*: 36–41.

Kundu, A.K., Lebedev, O.I., Volkova, N.E., Seikh, M.M., Caignaert, V., Cherepanov, V.A. and Raveau, B., 2015b. Quintuple perovskites $Ln_2Ba_3Fe_{5-x}Co_xO_{15-\delta}$ (Ln = Sm, Eu): nanoscale ordering and unconventional magnetism. *Journal of Materials Chemistry C*, *3*(21): 5398–5405.

Lagos, M.J., Trügler, A., Hohenester, U. and Batson, P.E., 2017. Mapping vibrational surface and bulk modes in a SingleNanocube. *Nature (London)*, *543*: 529.

Lebedev, O.I., Hébert, S., Roddatis, V., Martin, C., Turner, S., Krasheninnikov, A.V., Grin, Y. and Maignan, A., 2017. Revisiting hollandites: channels filling by Main-group elements together with transition metals in Bi2 – y v y V8O16. *Chemistry of Materials*, *29*(13): 5558–5565.

Lebedev, O.I., Turner, S., Caignaert, V., Cherepanov, V.A. and Raveau, B., 2016. Exceptional layered ordering of cobalt and iron in perovskites. *Chemistry of Materials*, *28*(9): 2907–2911.

Lebedev, O.I., Van Tendeloo, G., Amelinckx, S., Razavi, F. and Habermeier, H.U., 2001. Periodic microtwinning as a possible mechanism for the accommodation of the epitaxial film-substrate mismatch in the $La_{1-x}SrxMnO_3/SrTiO_3$ system. *Philosophical Magazine A*, *81*(4): 797–824.

Lekse, J.W., Natesakhawat, S., Alfonso, D. and Matranga, C., 2014. An experimental and computational investigation of the oxygen storage properties of $BaLnFe_2O_5 + \delta$ and $BaLnCo_2O_{5+\delta}$ (Ln = La, Y) perovskites. *Journal of Materials Chemistry A*, *2*(7): 2397–2404.

Lichte, H., 1993. Parameters for high-resolution electron holography. *Ultramicroscopy*, *51*(1–4): 15–20.

Lopatin, S., Ivanov, Y.P., Kosel, J., and Chuvilin., A., 2016. Multiscale differential phase contrast analysis with a unitary detector. *Ultramicroscopy*, *162*: 74–81.

Loretto, M.H., and Smallman, R.E., 1975. *Defect Analysis in Electron Microscopy*. New York: Springer.

Maignan, A., Martin, C., Pelloquin, D., Nguyen, N. and Raveau, B., 1999. Structural and magnetic studies of ordered oxygen-deficient perovskites $LnBaCo_2O_{5+\delta}$, closely related to the "112" structure. *J. Solid State Chemistry*, *142*: 247–260.

Moritomo, Y., Hanawa, M., Ohishi, Y., Kato, K., Takata, M., Kuriki, A., Nishibori, E., Sakata, M., Ohkoshi, S., Tokoro, H. and Hashimoto, K., 2003. Pressure-and photo-induced transformation into a metastable phase in RbMn [Fe (CN) 6]. *Physical Review B*, *68*(14): 144106.

Morniroli, J.P. and Steeds, J.W., 1992. Microdiffraction as a tool for crystal structure identificat ion and determination. *Ultramicroscopy*, *45*(2): 219.

Nelayah, J., Kociak, M., Stéphan, O., et al. 2007. Mapping surface plasmons on a single metallic nanoparticle. *Nature Physics*, *3*:348–353.

Nielsen, P.H. and Cowley, J.M., 1976. Surface imaging using diffracted electrons. *Surface Science*, *54*(2): 340–354.
Pennycook, S.J., 1992. Z-contrast transmission electron microscopy: direct atomic imaging of materials. *Annual Review of Materials Science*, *22*: 171–195.
Pennycook, S.J., and Nellist., P.D., 2011. *Scanning Transmission Electron Microscopy: Images and Analysis*, Dordrecht: Springer.
Rautama, E.L., Boullay, P., Kundu, A.K., Caignaert, V., Pralong, V., Karppinen, M. and Raveau, B., 2008. Cationic ordering and microstructural effects in the ferromagnetic perovskite $La_{0.5}Ba_{0.5}CoO_3$: impact upon magnetotransport properties. *Chemistry of Materials*, *20*(8): 2742–2750.
Raveau, B., and Seikh, M., 2012. *Cobalt Oxides: from Crystal Chemistry to Physics*. Berlin: Wiley-VSH Verlag GmbH .
Raynova-Schwarten, V., Massa, W. and Babel, D., 1997. Ein neues bleistrontiumferrat (111): die kristallstruktur der phase Pb4Sr2Fe60I5. *Zeitschrift für Anorganische und Allgemeine Chemie*, *623*: 1048–1054.
Rose, H., 1971a. Aplanatic electron lenses. *Optik*, *34*: 285–289.
Rose, H., 1971b. Properties of spherical corrected achromatic electron lenses. *Optik*, *33*: 1–24.
Rose, H., 1990. Outline of a spherically corrected semiaplanatic medium-voltage transmission electron microscope. *Optik*, *85*: 19–24.
Rosenauer, A. and Schowalter, M., 2008. STEMSIM—a new software tool for simulation of STEM HAADF Z-contrast imaging. In *Microscopy of Semiconducting Materials 2007, Springer Proceedings in Physics, 120*, ed. A.G.Cullis, and P.A. Midgley , 170–172 . Dordrecht: Springer.
Ruska, E., 1933. Formation of pictures of surfaces irradiated by electrons in the electron microscope. *Zeitschrift für Physik*, *83*: 492–497.497.
Scherzer, O., 1936. Über einige fehler von elektronenlinsen. *Zeitschrift für Physik*, *101*: 593–603.
Sheng, P., Li, W., Wang, X., Tong, X. and Cai, Q., 2014. Colloidal synthesis and photocatalytic performance of size-controllable solid or hollow $CuInSe_2$ nanocrystals. *ChemPlusChem*, *79*(12): 1785–1793.
Shindo, D. and Hiraga, K., 1998. *High-Resolution Electron Microscopy for Material Science*. Dordrecht: Springer-Verlag.
Sousa, V., Gonçalves, B.F., Franco, M., Ziouani, Y., González-Ballesteros, N., Fátima Cerqueira, M., Yannello, V., Kovnir, K., Lebedev, O.I. and Kolen'ko, Y.V., 2019. Superstructural ordering in hexagonal $CuInSe_2$ nanoparticles. *Chemistry of Materials*, *31*(1): 260–267.
Spence, J.C.H., 1981. *Experimental High-Resolution Electron Microscopy of Microscopy*. Oxford: Oxford Univ. Press.
Stuer, C., Van Landuyt, J., Bender, H., Rooyackers, R. and Badenes, G., 2001. The use of convergent beam electron diffraction for stress measurements in shallow trench isolation structures. *Materials Science in Semiconductor Processing*, *4*(1–3): 117–119.
Su, C., Tripathi, M., Yan, Q.B., Wang, Z., Zhang, Z., Hofer, C., Wang, H., Basile, L., Su, G., Dong, M. and Meyer, J.C., 2019. Engineering single-atom dynamics with electron irradiation. *Science Advances*, *5*(5): eaav2252.
Świerczek, K., 2011. Physico-chemical properties of $Ln_{0.5}a_{0.5}Co_{0.5}Fe_{0.5}O_{3-\delta}$ (Ln: La, Sm; A: Sr, Ba) cathode materials and their performance in electrolyte-supported intermediate temperature solid oxide fuel cell. *Journal of Power Sources*, *196*: 7110.
Tan, H., Verbeeck, J., Abakumov, A. and Van Tendeloo, G., 2012. Oxidation state and chemical shift investigation in transition metal oxides by EELS. *Ultramicroscopy*, *116*: 24–33.

Toda, A., Ikarashi, N. and Ono, H., 2000. Local lattice strain measurements in semiconductor devices by using convergent-beam electron diffraction. *Journal of Crystal Growth*, *210*(1–3): 341–345.

Turner, S., Verbeeck, J., Ramezanipour, F., Greedan, J.E., Van Tendeloo, G. and Botton, G.A., 2012. Atomic resolution coordination mapping in Ca2FeCoO5 brownmillerite by spatially resolved electron energy-loss spectroscopy. *Chemistry of Materials*, *24*(10): 1904–1909.

Van Aert, S., Batenburg, K.J., Rossell, M.D., Erni, R. and Van Tendeloo, G., 2011. Three-dimensional atomic imaging of crystalline nanoparticles. *Nature*, *470*(7334): 374.

Van Aert, S., Verbeeck, J., Erni, R., Bals, S., Luysberg, M., Van Dyck, D. and Van Tendeloo, G., 2009. Quantitative atomic resolution mapping using high-angle annular dark field scanning transmission electron microscopy. *Ultramicroscopy*, *109*(10): 1236–1244.

Van Tendeloo, G., Lebedev, O.I., Hervieu, M. and Raveau, B., 2004. Structure and microstructure of colossal magnetoresistant materials. *Reports on Progress in Physics*, *67*(8): 1315.

Verbeeck, J., Schattschneider, P. and Rosenauer, A., 2009. Image simulation of high resolution energy filtered TEM images. *Ultramicroscopy*, *109*(4): 350–360.

Verbeeck, J., Tian, H. and Schattschneider, P., 2010. Production and application of electron vortex beams. *Nature*, *467*: 301–304.

Vincent, R. and Midgley, P.A., 1994. Double conical beam-rocking system for measurement of integrated electron diffraction intensities. *Ultramicroscopy*, *53*(3): 271–82.

Völkl, E., Allard, F.L. and Joy, D.C., 1998. *Introduction to Electron Holography*. New York: Kluwer Academic/Plenum Publishers.

Volkova, N.E., Lebedev, O.I., Gavrilova, L.Y., Turner, S., Gauquelin, N., Seikh, M.M., Caignaert, V., Cherepanov, V.A., Raveau, B., and Van Tendeloo, G., 2014. Nanoscale ordering in oxygen deficient quintuple perovskite $Sm_{2-\varepsilon}Ba_{3+\varepsilon}Fe_5O_{15-\delta}$: implication for magnetism and oxygen stoichiometry. *Chemistry of Materials*, *26*(21): 6303–6310.

Wang, J., He, Y., Mordvinova, N.E., Lebedev, O.I., and Kovnir, K., 2018. The smaller the better: hosting trivalent rare-earth guests in Cu–P clathrate cages. *Chemistry*, *4*(6): 1465–1475.

Weyland, M., and Midgley, P.A., 2007. Electron Tomography In *Nanocharcterisation*, ed. A.I. Kirkland, and J.L. Hutchison, 184–267. Oxfordshire: RSC Publishing.

Wiktor, C., Meledina, M., Turner, S., Lebedev, O.I. and Fischer, R.A., 2017. Transmission electron microscopy on metal–organic frameworks–a review. *Journal of Materials Chemistry A*, *5*(29): 14969–14989.

Williams, D.B., and Carter, C.B., 1996. *Transmission Electron Microscopy A Textbook for Materials Science*. New York: Plenum Press.

Wu, Z., Yang, S. and Wu, W., 2016. Shape control of inorganic nanoparticles from solution. *Nanoscale*, *8*(3): 1237–1259.

Xu, L.C., Wang, R.Z., Liu, L.M., Chen, Y.P., Wei, X.L., Yan, H. and Lau, W.M., 2012. Wurtzite-type $CuInSe_2$ for high-performance solar cell absorber: ab initio exploration of the new phase structure. *Journal of Materials Chemistry*, *22*(40): 21662–21666.

Yarema, O., Bozyigit, D., Rousseau, I., Nowack, L., Yarema, M., Heiss, W. and Wood, V., 2013. Highly luminescent, size-and shape-tunable copper indium selenide based colloidal nanocrystals. *Chemistry of Materials*, *25*(18): 3753–3757.

Zewail, A.H. and Thomas, J.M., 2010. *4D Electron Microscopy: Imaging in Space and Time*. London: Imperial College Press.

Zhong, H., Wang, Z., Bovero, E., Lu, Z., Van Veggel, F.C. and Scholes, G.D., 2011. Colloidal $CuInSe_2$ nanocrystals in the quantum confinement regime: synthesis, optical properties, and electroluminescence. *The Journal of Physical Chemistry C*, *115*(25): 12396–12402.

4 Large Dataset Electron Diffraction Patterns for the Structural Analysis of Metallic Nanostructures

Arturo Ponce[1], *José Luis Reyes-Rodríguez*[1], *Eduardo Ortega*[1], *Prakash Parajuli*[1], *M. Mozammel Hoque*[1], *Azdiar A. Gazder*[2]

[1]Department of Physics and Astronomy, The University of Texas at San Antonio, San Antonio, Texas 78249, USA.

[2]Electron Microscopy Centre, University of Wollongong, New South Wales, 2500, Australia.

CONTENTS

4.1 Electron Diffraction Methods Under Transmission Electron Microscopy 112
4.1.1 Using the Microscope's Lens System 114
4.1.2 Using a Cs-Probe Corrector 114
4.1.3 Using External Hardware 115
4.2 Electron Diffraction of Sensitive Materials 115
4.2.1 Challenges in Electron Diffraction Used for Structural Determination of Proteins 116
4.2.2 Fast Scanning Nanobeam Electron Diffraction 119
4.3 Precession Electron Diffraction–Assisted Crystal Orientation Mapping 123
4.3.1 Correlation Contours for Twinned and Nontwinned Nanoparticles 125
4.3.2 Crystal Orientation in Hybrid Nanomaterials 126
4.3.3 Growth, Crystal Orientation, and Grain Misorientation in Thin Films 128
4.4 Electron Pair Distribution Function Applied to Protected Metallic Clusters 132
4.5 Concluding Remarks 138
Acknowledgments 139
References 140

4.1 ELECTRON DIFFRACTION METHODS UNDER TRANSMISSION ELECTRON MICROSCOPY

Structure of solids, including crystals, complex, structures, and amorphous materials, is mainly analyzed by ionizing radiation, which is classified into three groups: charge particles (electrons), photons (x-rays), and noncharged particles (neutrons). These different forms of radiation are accelerated at high energies, in ranges of thousands of electron-volts (keV). After the interaction with the solids a diffraction pattern (DP) is generated. In all of these types of radiation the diffraction is consistent with Bragg's law, which correlates the structure of solids from the direct to the reciprocal space, graphically represented by the Ewald sphere. Conventional x-ray diffraction (XRD) techniques are commonly used to study the structure of materials with different configurations such as bulk materials, thin films, powders, nanoparticles, and so forth. Additionally, the use of high-energy x-ray synchrotron sources, two diffraction methods are commonly employed neutrons and electrons. These two methods exhibit magnetic dipole moments that give rise to significant magnetic scattering in the analysis of materials. XRD and neutron diffraction (ND) are the standard techniques to investigate and resolve the crystalline structure of materials. Notwithstanding, these techniques rely on the interaction of at least micron-sized crystallites to produce a strong enough signal (Steuwer et al. 2004; Gu and Mildner 2016). Working over this size range could overlook fine interactions and mechanisms occurring at nanoscale (Pantzer et al. 2014; Ortega et al. 2017b; Betal et al. 2018). As a complement, electron diffraction techniques using the transmission electron microscope (TEM) allow the collection of specific analysis of individual phases, materials, and structures (Casallas-Moreno 2018; Legorreta-Flores et al. 2018; Ortega et al. 2018).

In electron microscopy (EM), the wavelength of electrons depends directly on the accelerating voltage of the source, which is normally employed by means of TEM. Electron diffraction coupled with TEM is the most used technique to characterize the structure of materials associated with imaging the specimen at the region from which the electron DP was taken. For several decades selective area electron diffraction (SAED) patterns have been used for the identification of crystalline phases, strain measurements, and the study of structural defects. The analysis of SAED patterns conventionally is performed in materials that are not affected by the beam irradiation or under external stimuli that can affect the structure. However, since in situ EM is becoming more popular, the collection of dynamic patterns is entering into a category of large datasets using ultrafast acquisition and high sensitivity under different illumination setups. Fast acquisition of electron DPs can be performed with a shutter that allows intermittent irradiation to reach the specimen or by means of scanning the sample under nanobeam TEM conditions. Another diffraction in EM is the electron backscatter diffraction (EBSD), which is coupled to a scanning electron microscope. EBSD is commonly used to study the structure of solids, for the analysis of crystalline phases and strain, and to obtain the crystal orientation mapping in the surface of materials. The DPs are collected after the electron-matter interaction at a high grazing angle on the surface of the sample, where the sample is tilted at about a 70-degree angle, giving rise to the formation of line patterns called

Kuikuchi patterns. In contrast, the electron diffraction modes in the optics of the TEM can be positioned in parallel or convergent illumination. In the former, closely related to TEM image mode, the intensity is procured from an area enclosed by a postspecimen aperture. To record significant data in a SAED pattern, the crystalline volume should be enough to overcome the signal produced by the amorphous support and other artifacts (Lábár 2005). In the latter illumination scheme, convergent beam electron diffraction (CBED), the electrons travel in a range of directions inside a cone, whose vertex is focused at the specimen. In this case, the CBED pattern has an array of bright disks used for structural and space group determination (Champness 1987). However, nanostructured materials either in 0D (nanoparticles), 1D (nanowires), or 2D (thin layers) require a combination of convergent and parallel illumination to reduce probe diameter. For those applications, microdiffraction or NBD is used to achieve crystallographic information on a point-by-point basis.

The electron diffraction under the illumination modes previously mentioned have been used extensively to study the structure of materials, to measure the strain, and to compare texture-related information of different size and ranges of materials (Matsui and Tabata 2012). (Sang, Kulovits and Wiezorek 2013) Electron diffraction in TEM has several advantages compared with analogue diffraction methods, neutrons, and x-rays. One important advantage is the use of lenses to produce an image of the analyzed region at subnanometric resolution. In fact, the imaging methods in TEM are directly related to the diffraction, for instance, the diffraction contrast mode can produce bight-field (BF) and dark-field (DF) images. By tilting the specimen and exciting a reflection in the DP, one can reach the two-beam conditions to characterize structural defects such as dislocations, grain boundaries, stacking faults, and precipitations. In this way, electron diffraction combined with the power of magnification of the microscope has the advantage that is not found in other diffraction method. Convectional electron DPs are registered without alteration of the specimen, if the material is not sensitive to be damaged by the high energy of the incident beam. When the sample is stable the electron pattern does not change over time, the electron beam interacts permanently with the specimen and can be considered static because the deflector coils in the column of the microscope do not deflect the beam. Under these conditions, the electron beam is not called "dynamic" if full automation is not performed; therefore, the data collection, indexing, and graphic interpretation must be performed separately. The results are postanalyzed by using the crystallographic database by modeling and simulating the structure using computational tools that finally will be correlated with the experimental data collected. An automatic analysis and data collection are necessary for special requirements of the specimen. The collection of large dataset electron DPs arises from the special requirements of the specimen, such as the reduction of the radiation damage, the analysis of crystals in a polycrystalline material, and the characterization of individual nanoparticles dispersed in a film. A fully reproducible and systematic analysis over wide ranges would require a traceable scanning probe to allow pixel correlation. Over the past years, a combination of software and instrumentation has been developed to fulfill this objective (Darbal et al. 2013; Viladot et al. 2013). These approaches can be divided into three categories: (1) using the microscope lenses, (2) using a spherical-aberration-corrected system (Cs-Corr), and (3) using additional hardware.

4.1.1 Using the Microscope's Lens System

With the introduction of field emission guns (FEGs), high intensity, sensitivity, and coherent electron DPs were routinely produced (Cowley 1999). It was proven by Cowley (1996) that by combining the traditional convergent beam technique and optimum defocus of the electron beam a 2-nm probe size could be achieved. Their methodology relied heavily on overexciting the voltage values of the prespecimen lenses to optimize the beam convergence angle, giving more flexibility to the illumination system. Examples of diffraction scanning transmission electron microscopy (D-STEM) (Figure 4.1a) technique have been reported for the characterization of nanomaterials and biochemical systems (Cowley et al. 2000; Quintana, Cowley and Marhic 2004). Using the full set of condenser lenses (CL1, CL2, CL3, and Lorentz lens, CM), Ganesh et al. (2010) achieved an electron probe, with parallel illumination, under the 2-nm range. Using BF and DF images as references, it was possible to position the electron probe directly into the region of interest to perform the analysis (Darbal et al. 2013).

4.1.2 Using a Cs-probe Corrector

Microscope instabilities, parasitic aberrations, and optical aberrations limit the resolution achievable during normal TEM imaging and spectroscopic acquisition. The most significant of these optical aberrations, spherical aberration (Cs), requires a special set of lenses and hardware to reduce it (Hawkes 2015). This additional set of coils can be used to manipulate the electron beam. D-STEM patterns can be collected by positioning a raster probe using the Digiscan control for subsequent acquisition with a charge-coupled device (CCD) camera (Zankel et al. 2009). The overlapping of the convergent disks is achieved by modulating the second CL (CL2) and the adaptor lenses (ADLs) of hexapole coils of a Cs corrector (Figure 4.1b). Fine spot alignment is completed using the shift and beam deflectors as a way to provide a beam as convergent as possible (Tlahuice-Flores et al. 2013). This setup was used to characterize the radiation-sensitive Au_{130} and Au_{144} clusters. Under these conditions an electron probe of 0.09 nm with a current of 22 pA can be used to collect local data.

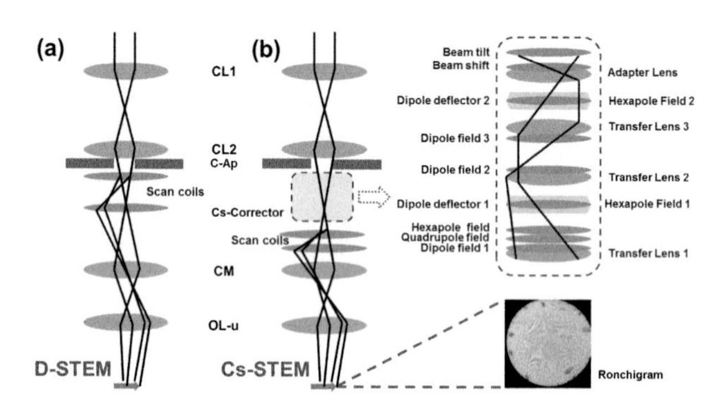

FIGURE 4.1 Ray diagrams for (a) D-STEM and (b) aberration-corrected STEM (Cs-STEM). **Reprinted with permission from (Ortega et al. 2018), copyright (2018), ProQuest LLC.**

Due to the high electron dosage, the Gold-thiolated nanoparticles exposed to continuous irradiation are damaged permanently. Therefore, the relevant information is collected from the first milliseconds of the particle-beam interaction (Bahena et al. 2013; Tlahuice-Flores et al. 2013).

4.1.3 Using External Hardware

Commercial instrumentation, used to perform precession electron diffraction (PED) or crystal orientation mapping, can be used to control the different sets of coils of the microscope (Moeck and Rouvimov 2010). In the PED method, the electron diffraction is registered while the electron beam is precessing on a cone surface; in this way, only a few reflections are simultaneously excited at the same time, therefore, dynamical effects are strongly reduced. For this type of setting, the scan and descan coils need to be calibrated to maintain the beam movement on range so that electron DPs remain stationary during precession and scanning routines. Because these systems use external cameras collecting data from the inclined focus screen of the microscope, camera length and distortions need to be corrected during data treatment. The collected DP areas are then cross-correlated from database templates of the crystal structures present in the sample (Santiago et al. 2016). The commercial use of software like ASTAR has found a range of applications going from metallurgy to in situ heating of experiments inside the TEM (Rauch and Veron 2005; Kobler et al. 2013; Deng et al. 2015).

4.2 ELECTRON DIFFRACTION OF SENSITIVE MATERIALS

Low-dose, low-voltage (80 kV), and cryo-TEM are some approaches implemented to reduce radiation damage. The efficacy of these methods relies on reducing the electron density or the exposition time. Historically, CCD cameras and photographic film have played important roles as recording media. Acquiring, visualizing, and analyzing images at a very high frame rate, without sample deterioration, has been a major challenge in TEM. The current technology for image acquisition uses high electron doses to collect images to guarantee a high signal-to-noise ratio and high-speed data acquisition. In situ TEM permits the manipulation and stimulation of samples while they are imaged, record the spectra, or to measure what changes are taking place in real time, in which a fast data collection is crucial. In situ TEM experiments included in the proposal are in situ heating and electrical biasing and in situ TEM imaging with fluid cells. Heating/electrical biasing will be integrated with functional electrode layouts for electrical biasing experiments of metallic nanostructures and semiconductor thin films. In situ liquid cells will be used to study the diffusion and self-assembly of nanoparticles in confined liquid environments. To record reliable electron diffraction data, there are two limitations to overcome: reduction of dynamical effects due to the multiple scattering and the saturation and blooming effects suffered by recording media due to the high intensity of the direct beam when compared with the nearest reflections (Figure 4.2) (Rauch et al. 2010). Quasi-kinematical DPs can now be routinely recorded with the use of PED, and blooming

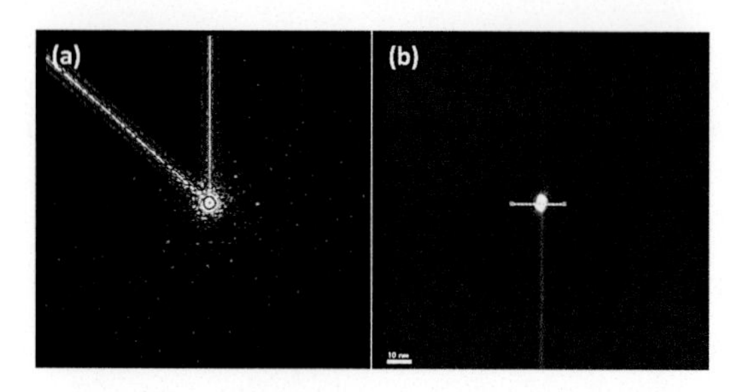

FIGURE 4.2 Comparative patterns taken with cameras. (a) A CCD (blooming effect due to the saturation of the camera) and (b) CMOS (no blooming effect is presented).

effects (overflown of the charge accumulation of the CCD photoactive region into the neighbor capacitive well) have been diminished with the use of highly sensitive complementary metal oxide semiconductor (CMOS) active pixel detectors, which are capable of distinguishing even single electron events (Moeck et al. 2011; Allé et al. 2016). Therefore, a critical compromise between damage and signal-to-noise ratio needs to be considered. For this reason, techniques as fast scanning nanobeam diffraction and pair distribution function (introduced later on in this chapter), provide an efficient way to characterize sensitive small clusters without deterioration of its structure.

Modern direct electron (DE) detectors enable the study of dynamic processes for in situ EM recorded over a range of resolution and speed combinations with a full-frame streaming at up to 6000 frames per second (fps). In this way, the streaming video technology, which uses postevent trigger video capture, allows the capture of in situ TEM experiments from the beginning and during the experiment. Additionally, the DE detectors are optimized to collect large fields of view with ultrahigh sensitivity under low-dose conditions and cryo-EM for studying the structure of samples susceptible to damage for imaging and electron diffraction modes.

4.2.1 Challenges in Electron Diffraction Used for Structural Determination of Proteins

The described optimal configuration of modern cryo-electron microscopes has pushed the method to near-atomic resolution, with several protein structures already solved under 3-Å resolution. Cryo-EM has had such a big impact in the field of structural biology that *Nature Methods* designated it as Method of the Year 2015 (2016). There are two avenues in structural biology to determine the structure of proteins: single particle technique is performed at high resolution and micro-electron diffraction (MicroED). The main damage mechanism that can appear in proteins during the acquisition of images or electron diffraction in the TEM is the ionization damage (radiolysis). Certain conditions must be satisfied to declare that the observed electron DPs are significant for the case of proteins susceptible to beam damage. They must show stability for several milliseconds, i.e. they are the representative patterns registered under rapid acquisition and ultralow-dose irradiation, to preserve not only the initial structure of the protein. In this way, brightness is directly related to the electron current density per unit solid angle of the source. The current density can be measured from the fluorescence screen to convert it to the dose rate on the specimen. However, to determine the dose rate the magnification factor, which is also proportional to the radius of the viewing screen, needs to be considered. In the microscope, this magnification is referred to as the film plane, although there is no film the magnifications still use that plane as a reference. Hence, the magnification of the images recorded using a camera correspond to around 80% of that on the film plane. Therefore, we can relate the current density on the screen to that on the specimen by the following formula (Ortega et al. 2017a):

$$\sigma = \rho \times C \times (0.8M)^2 \tag{1}$$

where σ is the number of electrons per area ($\overline{e}$ /Å^2), ρ is the current density measured on the screen viewing (pascal per square centimeter), M is the magnification on the screen, and C ($\sim 6.25 \times 10^{-10}\ \overline{e}\ \text{cm}^2\ pA^{-1}\ \text{Å}^{-2}$) is a proportionality constant to relate the current per square centimeter to electrons per square Angstrom. These units need to be converted properly considering that the dose or dose rate is calculated within the period of certain exposure time in which the shutter is open to irradiate the sample. Conventional high-resolution TEM (HRTEM) is around 100 pA/cm^2, which represents thousands of electrons per square Angstrom due to the large magnification. If we consider radiolysis as the primary radiation damage in the proteins, then for an HRTEM image taken at $M = 500$ K the electron dose exerted on the sample will be around 12,500 $\overline{e}$ /Å^2 s. This dose rate is capable of producing changes in the structure of the proteins.

Several methods are currently used to determine the structure of proteins, including x-ray crystallography, nuclear magnetic resonance (NMR) spectroscopy, and EM. Although the current databases for protein structures are mainly based on the first two methods, EM is gaining use as an alternative method to study proteins that cannot not be analyzed by x-ray or NMR. However, the necessity of good crystallization and sufficient size remain as the main disadvantages of this method, because some proteins, like those located in cellular membranes, cannot be crystalized. Although traditional EM provides important information about proteins, there are some intrinsic limitations in the method that restrict the resolution of the information. First, the samples have to be introduced in a high vacuum requiring the removal of the natural aqueous environment or a chemical conservation. This procedure limits the kind of specimens we can study and the accuracy of the information we can retrieve. For instance, liposomes, micelles, polymers, or even proteins disintegrate or change their shape under dehydration. Second, the low-contrast nature of biological samples requires the introduction of contrast agents. However, the staining process has been reported to introduce artifacts or interact with samples, changing the true nature of the original specimens. Third, the interaction of the electron beam with the light atoms of proteins rapidly affects the samples through ionization and breaking of chemical bonds, resulting in heating of the samples and destroying the crystallinity in cases such as protein crystals. These phenomena require the use of low-dose techniques or lower electron acceleration voltages to the detriment of the resolution of the image. These three challenges in the EM observation of proteins have been addressed by the implementation of the cryo-EM (cryo-TEM). In cryo-TEM, the fully hydrated specimens are ultra-rapid cooled in liquid ethane to prevent ice crystal formation and then observed in a cryo-microscope at liquid nitrogen temperatures. The information obtained in this cryo-fixed state is closer to the native state of the specimen. The rapid freezing of the sample in its original aqueous environment avoids the first challenge of the traditional TEM, such as ultrastructural changes, agglomeration caused by water removal, or washing away of substances by chemical processing. As samples are generally frozen in a natural state without any contrast agent, the cryo-TEM avoids the second challenge, the generation of artifacts by staining agents. Finally, the low temperature of the cryo-TEM partially avoids the third challenge, because the low diffusion of broken fragments at low temperature results in less conformational changes, slightly increasing resistance to radiation damage. In contrast, modern cryo-electron microscopes are optimized for low-dose procedures as well as enhancement of contrast. Biological samples must be imaged with a small total dose (20–30 electrons/$Å^2$), as higher accumulated dose degrades the specimen. Last-generation cryo-electron microscopes include smaller condenser apertures to reduce the dose on the sample and a special lens condenser system to ensure parallel sample illumination over a wide and variable field of view. In addition, the low-dose conditions require that the area to be imaged is not exposed before the actual picture is taken. This can be achieved by focusing on an adjacent area of the sample and then quickly moving to an unexposed area to record the images. In this direction, most cryo-electron microscopes incorporate low-dose software designed to minimize electron dose during operation, as well as a high degree of automation in data collection. On the other hand, the low signal-to-noise ratio of the samples (low contrast) implies that many images must be averaged to suppress the noise and amplify the signal (the structure of the specimen). However, movement of a particle over the course of the exposure will lead to signal degradation. For MicroED, the DE camera allows for fast read-out for dynamic experiments. The high speed and high sensitivity is ideal to study sensitive samples, such as proteins, cells, enzymes, and so forth, before they are damaged. The structural determination of a crystallized protein is obtained by quantifying the intensities of the reflections of DPs, ideally collected under PED conditions and combined with 3D tomography. The 3D tomography in imaging is a process that collects a series of 2D images at different angles; the experimental data are subsequently processed to obtain a reconstruction and 3D visualization. The data collection of electron DPs in a protein is performed automatically at high angles; for example, if the range is 120°, then 120 DPs are registered with a step of 1°. The crystal cell in the reciprocal space can be reconstructed, and the crystal cell parameters can be obtained with a high accuracy (error between 2 and 5%), this method is known as the automated electron diffraction tomography (ADT) (Kolb, Mugnaioli and Gorelik 2011).

4.2.2 Fast Scanning Nanobeam Electron Diffraction

This section addresses the methodologies used to study the structure of thiolated metallic clusters by using fast scanning nanobeam diffraction to register electron DPs in fractions of seconds. The rapid scanning of the high energetic beam interacting with the metallic clusters induces an angular momentum due to the drift velocity of electrons through the sample. The adequate control of shape, size, and crystal structure of nanoparticles constitutes the fundamental challenge to overcome in the current state of the art of nanotechnology to develop nanomaterials with practical applications. The ability to image specimens in their native aqueous state is a powerful tool not exclusive to structural biology. Cryo-microscopy is increasingly becoming an interdisciplinary tool over a broader range of fields. A prominent example is the characterization of colloidal nanoparticles and nanocomposites such as the ligand-protected metallic clusters. In the ligand-protected metallic clusters two simultaneous key damage mechanisms appear: (1) the knock-on effect, occurring due to the momentum transfer between electrons and atom cores, and (2) radiolysis (ionization damage) effects, which degrade the p-mercaptobenzoic acid ligands and as a result the stability of the cluster (Egerton 2019). In the case of organic solids, the damage diminishes with increased accelerating voltages, decreased beam currents, and reduced exposure (Egerton 2013). To address these issues the combination of low electron dosage and fast acquisition is required, which is described in this section.

Conventional HRTEM and SAED are typically obtained with a large condenser aperture (30–40 μm), which can exert an electron density of around 100 pA/cm^2 over the viewing screen of the TEM. For an HRTEM image taken at a magnification of $M = 500$ K the electron dose can reach around 12,500 $\overline{e}$ /Å^2 (Ortega et al. 2017). This dosage is capable of producing irreversible changes in the structure of small metallic clusters. The images shown in Figure 4.3a–d highlight the structural changes from an individual the ligand-protected metallic cluster $Au_{144}(SR)_{60}$, which exhibits a structural transformation that goes from a face-centered cubic (fcc) structure to a decahedral particle oriented along the fivefold symmetry during continuous irradiation. The sequence of images in Figure 4.3a–d were registered at 200 ms using a 2K × 2K CMOS camera.

Under nanoprobe mode, with a saturated CL and the smallest condenser aperture (5 μm), the sample can be irradiated in a quasi-parallel illumination. The single spot produced by the focused beam is then able to probe individual nanoparticles. Unfortunately, as shown in the sequence of images in Figure 4.3e–h, as the current density increases, the irradiated area decreases. To register high-quality diffraction data without damaging the clusters, the patterns need to be recorded at a fast rate.

The raster movement of the electron beam can be possible due to the manipulation of a conventional TEM lens system controlled by a PED-assisted automated crystal orientation mapping (PED ACOM-TEM) unit connected to the microscope (Shigesato and Rauch 2007). In the standard ACOM-TEM technique, the probe is scanned across the area of interest and collects an electron DP per step/pixel using an ultrafast CCD camera attached to the viewing screen of the microscope (Ruiz-Zepeda et al. 2014). However, CCD architecture suffers several drawbacks that limit

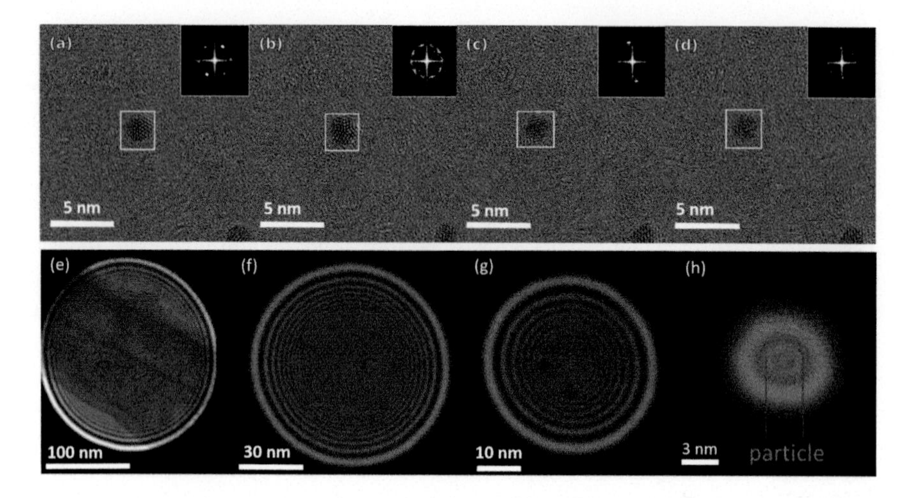

FIGURE 4.3 Frame-shot sequence of $Au_{144}(SCH_2CH_2Ph)_{60}$ on amorphous carbon. The insets show the fast Fourier transform for the framed region. The structure of the particle is modified by the irradiation: (a) fcc-like orientation, (b) fivefold orientation, and (c) and (d) two different orientations. Irradiated areas in nanoprobe mode are (e) 22 $\overline{e}$ /Å^2, (f) 65 $\overline{e}$ /Å^2, (g) 250 $\overline{e}$ /Å^2, and (h) 1400 $\overline{e}$ /Å^2.

Reprinted with permission from (Ortega et al. 2017a), copyright (2016), Springer.

fast-frame acquisition: serial data access (slow for large arrays), high power dissipation (on the range of CV^2f), and large voltage drive levels for charge transfer (Magnan 2003). For this reason, to allow the collection of the weak reflections produced by the small crystalline volume of a single cluster, the scan size and movement is synchronized with the microscope's in-line camera. On this example, the DPs of $Au_{102}(MBA)_{44}$ clusters are recorded with a Tietz Video and Image Processing System (TVIPS) 16-mega pixel CMOS camera with a dynamic range (maximum/noise) of 10,000:1 (TVIPS-GmbH TemCam XF-Series 2019). A schematic representation of the experimental setup is illustrated in Figure 4.4. Under this configuration, the probe size is around 2 nm. The ACOM-TEM controlling unit allows for automatic tilting and subsequently de-scanning of the electron beam using the image shift coils of a standard TEM [not originally intended for scanning TEM (STEM)]. The resulting DP appears as a sequence of stationary spot images while the scanning is performed line by line over the field of view previously selected. The patterns are recorded in video mode at a frame rate of 10 Hz with a beam current of 2 pA/cm^2 and a probe size of 2 nm bigger than the 1.6 nm $Au_{102}(MBA)_{44}$ clusters. Different areas of 50 × 50 nm^2 at 250 Kx were scanned. Under these settings the exerted dose rate per pattern accrued to 1350 $\overline{e}$ /Å^2 (Ortega et al. 2017a). The individual images are subsequently processed and compared with simulated patterns of the theoretical structure. The simulated DPs have been generated by using the *xyz* cluster coordinates provided by XRD of microsize crystalized clusters.

The electron diffraction simulations were performed using the module "Nanodiffraction" in the Java Electron Microscopy Software (JEMS)

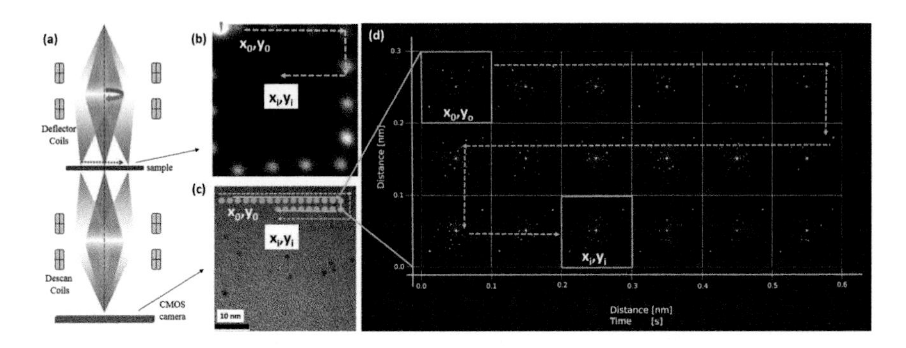

FIGURE 4.4 (a) Visualization of the experimental setup on the JEOL 2010F that shows the path the electron beam takes after the deflector coils, over the sample, and into the CMOS camera. (b) TEM image of the region of interest showing dispersed nanoparticles. (c) Prolonged exposure of the scanning area (squared perimeter) used to adjust the scan settings. (d) Nanobeam electron diffraction (NBD) patterns collected on a single shot; the set of images are acquired from a 3 × 6-Å² subarea.

(Stadelmann 1987). For the digital comparison, the cluster is simulated over an orthogonal supercell of 36 × 36 × 18 Å to avoid periodicity effects. To match the expected diffraction data with the experimental data it is important to consider that the particle can rotate over its own axis and produce a different projection. Taking the fivefold symmetry structure as the origin of the *z*-orientation in the coordinate system, the nanoparticle is rotated on 1-degree steps in five different directions—*x*, *y*, *xy*, ½*xy*, *x*½*y*—to obtain a dataset of 1800 simulated DPs in which the main fitting parameter is the relationship between spot positions and angles (Figure 4.5a) (Ortega et al. 2016).

To overcome the amorphous carbon signal, image processing is needed to pinpoint the key features hidden in the experimental DPs. A filtering procedure is applied to all images for consistency. The postprocessing is done applying a Lucy-Richardson filter and then a Gaussian filter based on application differences to deconvolute and

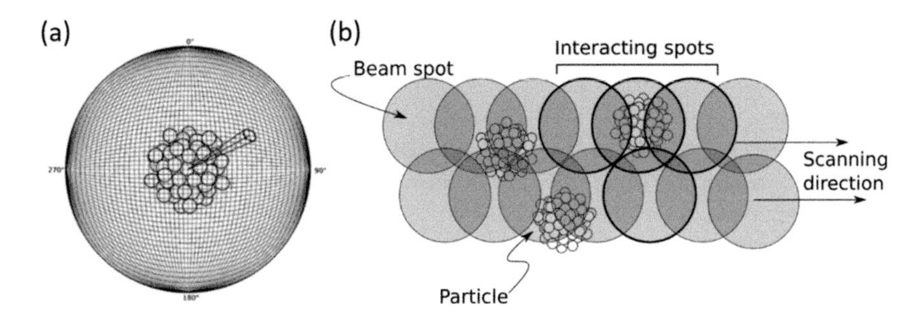

FIGURE 4.5 (a) Schematic representation of the theoretical data bank of the simulated $Au_{102}(MBA)_{44}$ DPs used for comparison. (b) Typical line-by-line raster scanning of the electron probe, as the beam overlaps it can interact with the same nanoparticle multiple times over slightly different "crystalline centers."

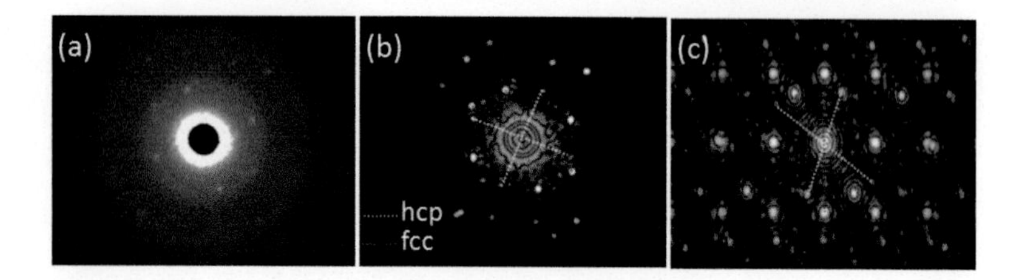

FIGURE 4.6 (a) Example of an experimental DP from A of the $Au_{102}(MBA)_{44}$ cluster. (b) A filtered DP to enhance contrast, two structures were considered, face centered cubic (fcc) and hexagonal close packed (hcp). (c) The corresponding simulated pattern.

enhance the signals (Fish et al. 1995). Figure 4.6b shows the resulting postprocessed pattern to compare with the simulated set as the one depicted in Figure 4.6c. Both filters are implemented in the software G'MIC using its command line interface for batch processing (Tschumperlé and Fourey 2016). The choice of initial parameters was tested with the GIMP software (www.gimp.org). The resulting image only preserves the relative positions as the original ratio of intensities is lost after filtering. Therefore, the matching is based on a quantitative comparison between spot positions and angles between the experimental and the simulated DPs (Bahena et al. 2013). These structures (given by those coordinates) are stable only when they are protected by ligands. In the simulation of DPs, the pMBA ligands are removed "artificially" while keeping the positions and geometry of the metallic atoms. The removal of these organic ligands is performed to reduce background noise and enhance the intensity of the reflections for comparison with the experimental ones. It is important to mention that additional errors in angle measurements may occur following the image-processing algorithm. The flatten intensities and the signal enhancement (increased image sharpness) can induce small spot-centering shifts. These variations are by-products of the blind image deblurring algorithms used, which estimates an associated blur kernel to solve the images' noise-removal minimization problem. Deviations from the computational blur kernel and the CMOS camera response will introduce dispersion of the data (Bruma et al. 2016).

Depending on pixel and area size, a full scan can comprise thousands of individual frames. Each frame represents a DP (or the beam image in the case of probing the carbon film) that corresponds to a specific "pixel" of the scan area. Due to the beam probe size and the scan spacing, a single particle can interact with the electron beam several times (Figure 4.5b); these DPs possess certain similarities but are not equal (Figure 4.7). The patterns shown in Figure 4.7 correspond to the same nanoparticle but are "probed" 1° away. This means the electron beam interacts and tilts the cluster as it scans the area and "touches" it up, down, left, and right, as shown in Figure 4.7a–e (Ortega et al. 2017a). From the DPs it is observed that the particle is oriented almost in the same direction during scanning, i.e. the beam-particle interaction is such that it does not significantly (~1°) disturb the orientation of the $Au_{102}(MBA)_{44}$ cluster.

Under the conditions described, the clusters preserve their structure. Herein, the stability of each DP after multiple "beam passes" is used to assess the structural

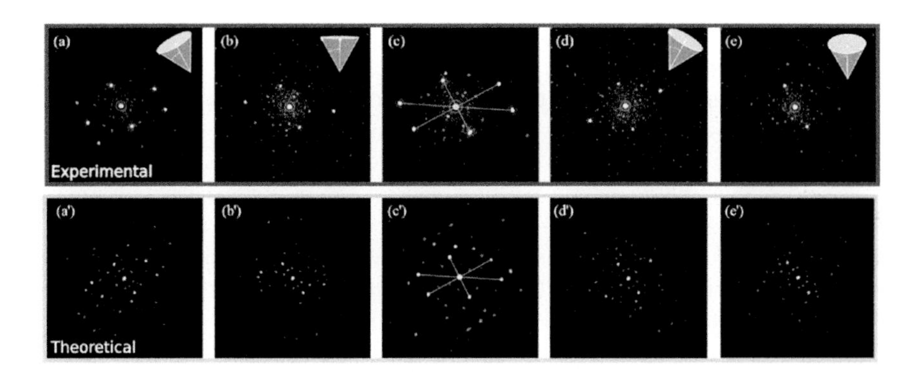

FIGURE 4.7 Set of experimental fast-NBD and simulated patterns extracted from the $Au_{102}(p\text{-}MBA)_{44}$ cluster. From an arbitrary orientation of the nanoparticle (c), a conical oscillation of the cluster is observed from the surrounding DPs (a, b, d, and e). These 1-degree variations correspond to a "left, up, right, and down" tilting. Diffraction patterns a'-e' were simulated and used to index the experimental ones to

Reprinted with permission from (Ortega et al. 2017a), copyright (2017), Springer.

integrity of the cluster. Although the electron doses of this method match conventional diffraction techniques, the advantage of the nanobeam electron diffraction (NBD) setup relies on the acquisition of specific local media as opposed to SAED, where we collect the superimposed information of all the enclosed area. This approach is sensitive enough to even differentiate small angle rotations produced by the interaction of the clusters with the electron probe.

4.3 PRECESSION ELECTRON DIFFRACTION–ASSISTED CRYSTAL ORIENTATION MAPPING

PED is a diffraction technique for structure determination, developed by Paul Midgley and Roger Vincent (Vincent and Midgley 1994) to measure the diffracted intensities achieving near kinematical conditions in the DP. In PED, the beam tilt coils are used to get a conical swing of the electron beam around a pivot point in the sample plane. Afterward, the image shift coils are used to de-scan the electron beam to obtain a quasi-kinematic DP in the diffraction plane. The electron beam can be convergent or parallel, and it is inclined away from the optical axis forming a precession angle (φ) with typical tilts between 0 and 3°. During the precession of the electron beam, the precession frequency is variable, but generally it is set to 50 Hz or 100 Hz. Reflections are scanned at each moment by the Ewald sphere with only a few reflections under the Bragg condition, which promotes the accrual of a DP with dynamic effects less pronounced, representing the sum of a continuous set of electron DPs (Zou, Hovmöller and Oleynikov 2011). Diffraction spots maintain a distribution of intensities in a proportional ratio closer to the square of the structure factor; however, the multiple scattering and the prohibited reflections are considerably reduced. An increment in the precession angle will allow that Ewald sphere to touch more points in the reciprocal space, thus more diffraction points will be shown in the DP. Therefore, the PED technique allows the acquisition a more refined structure analysis.

Figure 4.8a shows a schematic representation of the PED-assisted crystal phase. The incident electron beam can be tilted and rotated in a conical hollow surface around the optical axis using the beam tilt coils. The diffracted intensities are descanned with the image shift coils, and then the DP is displayed as a stationary spot pattern. An external ultrafast CCD camera, which is attached to the viewing screen of the microscope, registers every step of the scanned area and records each electron DP to be stored in a computer for a subsequent indexing process. The pattern recognition process is performed by a comparison between calculated cross-correlation values. For each experimental pattern, there is one value for a theoretical DP template computed for every possible orientation and for every expected phase through the crystal lattice parameters from the crystallography open database (COD). Figure 4.8b shows five electron DPs from their respective region in a gold nanoparticle. All these patterns are registered close to the [110] zone axis. The corresponding simulated electron DP is also shown in Figure 4.8c.

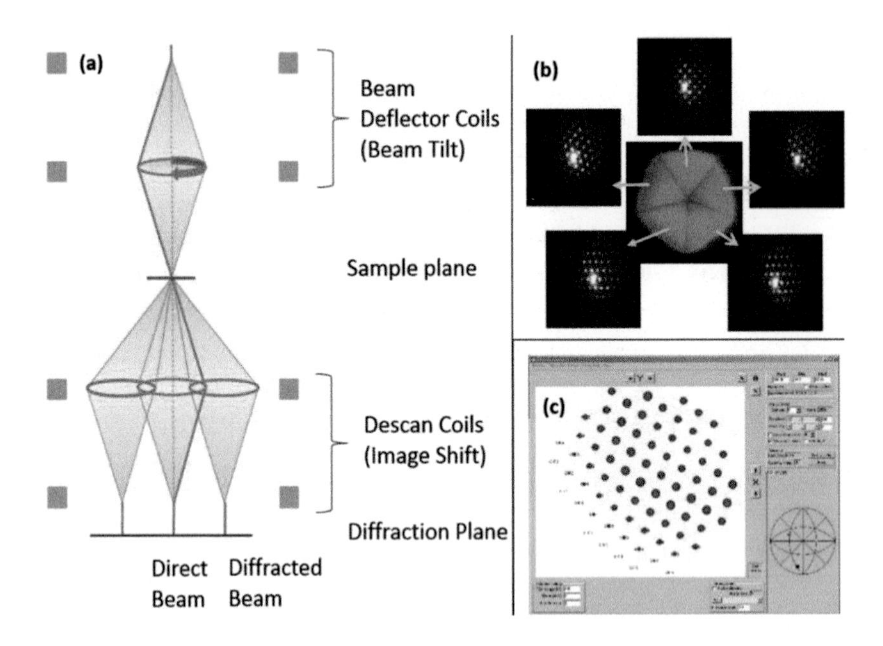

FIGURE 4.8 (a) Schematic representation of the e-beam in the column of the microscope under a precession geometry. (b) Individual electron DPs taken in five regions along the [110] direction. (c) Example of one simulated electron DP used for matching with the experimental ones.

4.3.1 Correlation Contours for Twinned and Nontwinned Nanoparticles

In this section, the crystalline orientation mapping applied to polyhedral nanoparticles is employed to multiple twinned metallic nanoparticles. Nanoparticles with different shapes can be obtained as a consequence of the balance of the binding energies of atoms into the nanoparticle and those located at surface planes during the growth stage. The use of surfactant additives during the synthesis of nanoparticles promotes their stabilization and enables the modification of the growth rate and the surface free energy; the latter plays a crucial role in the final shape of the particles (Baletto and Ferando 2005; Barnard 2014). Stability of nanoparticles depends mainly on their size range and by faceting. Smaller clusters of a few atoms to particles of a few nanometers are very unstable if they are not stabilized in a certain media. In contrast, highly faceting structures such as octahedral, decahedral, or icosahedral (Zhou et al. 2008; Carbó-Argibay et al. 2010) (Platonic solids) nanoparticles are very stable and over a size range higher than 50 nm maintain a mostly polyhedral morphology based on Wulff-Ringe (Ino 1969; Marks, 1984; Ringe, Van Duyne and Marks 2013) construction formalism, which is applied to multiply twinned particles and oxide nanoparticles. The use of a small and focused electron probe is useful to characterize nanoparticles of this size range (~100 nm) as the point-by-point data acquisition of the PED ACOM-TEM technique allows for correlation and quantification of the crystalline changes during scanning. This permits the visualization and analysis of very complex structures that can be produced by faceting. This type of feature would be easily ignored by standard TEM imaging techniques or volumetric diffraction methods. As an example, the next images investigate the shape of a modified truncated decahedron, which exhibits a barrel-like shape with multiple facets by PED.

Figure 4.9 a–h shows selected gold nanoparticles to highlight the presence of twin boundaries. The set of electron DPs was obtained by scanning the precessed electron beam over the region of interest. The DiGISTAR precession unit from NanoMEGAS was operating at 50 Hz with a precession angle of 0.9°. NBD conditions were used in the JEOL ARM 200F microscope to achieve a probe smaller than 2 nm. The camera length and the distortions introduced by the position of the external CCD camera were corrected during data treatment. Figure 4.9a–d shows regular decahedral particles in which a virtual BF and contour images of several nanoparticles with twin planes are visible. To reconstruct a virtual BF image from the DPs, a virtual aperture is placed over the transmitted beam of the recorded DPs. Then, the average intensity is calculated leaving out the contrast information from the diffraction spots. In this sense, the image formed is a STEM type BF image of the object. Figure 4.9c and d shows the corresponding correlation contour maps exhibiting clearly inner contrasts that correspond to the multiple twins' boundaries contained within the nanoparticles. Contour maps are built cross-correlating the pixels/DPs from the set of first neighbors. Identical DPs in the vicinity will produce a white pixel to denote the continuity of the same crystalline orientation or phase. In contrast, a difference in the surrounding pixels will be indicated by a grayscale value. As an example, the pinpointed nontwinned particles, denoted by arrows in the larger scan in Figure 4.9g and h, show contours at their perimeter. Because the difference between the crystalline DP and amorphous carbon is high, the corresponding cross-correlated pixel color is black.

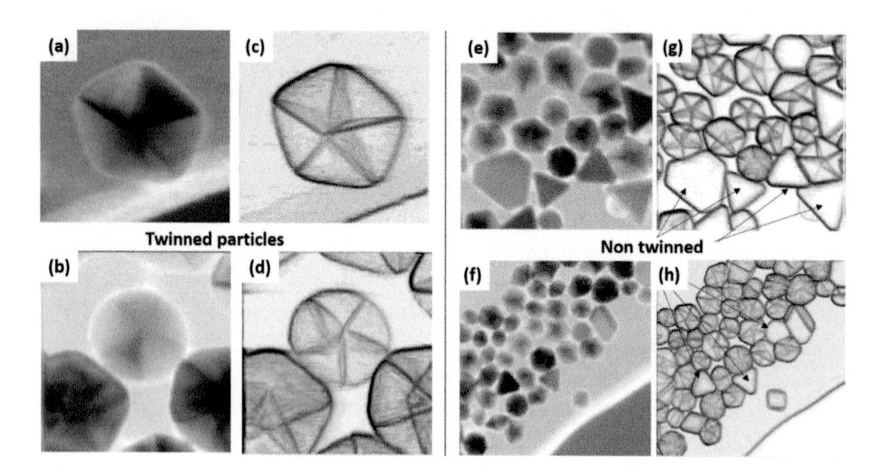

FIGURE 4.9 Virtual bright-field (VBF) and contour images obtained by PED ACOM-TEM of decahedral nanoparticles. (a and b) Show the signal derived from the integration of the center beam, which is equivalent to a standard BF image. (c and d) Corresponding contour image obtained cross-correlating the first set of neighbor pixels. The contour map highlights the presence of twin planes and/or changes in crystallinity or phase. (e–h) Larger scans showing the presence of nontwinned nanoparticles (pointed by arrows).

Reprinted with permission from (Santiago 2016), copyright (2019), Elsevier.

4.3.2 Crystal Orientation in Hybrid Nanomaterials

Hybrid materials at nanoscale combines the physical properties of materials to create smart and tunable materials for specific application. For example, in transition metals magnetic, plasmonic, and catalytic properties are well studied in metals such as cobalt, silver, and platinum, respectively. Particularly, the magneto-optical interactions have gained attention due to the potential technological applications in spintronic, electromagnetic shielding, magneto-optical data storage, and others (Maksymov 2015, 2016). Nanoscale magnetic structures, and their ordered arrays, are being considered for use in several advanced technological areas such as microelectromechanical systems (MEMS) and power devices such as supercapacitors or batteries. Metal-semiconductor materials have attracted great interest in the fabrication of nanoantennas, where the coupling of metal with semiconductors leads to the development of resonant nanostructures responding to a specific frequency bandwidth. In this context, zinc oxide has been recognized as an excellent semiconductor material in which its morphological configurations play an important role in its properties and applications (Sanchez et al. 2016). The crystalline structure plays a determinant role in the properties due to the crystalline anisotropy as well as the heterojunction of the materials. The hybrid material is a ZnO/Ag nanoantenna, where the core is a silver nanowire with a pentagonal cross-sectional area. The five surfaces of the ZnO rods are assembled using microwave synthesis (Sanchez et al. 2017). The crystallographic analysis reveals five nucleation sites for the ZnO nanorods, which

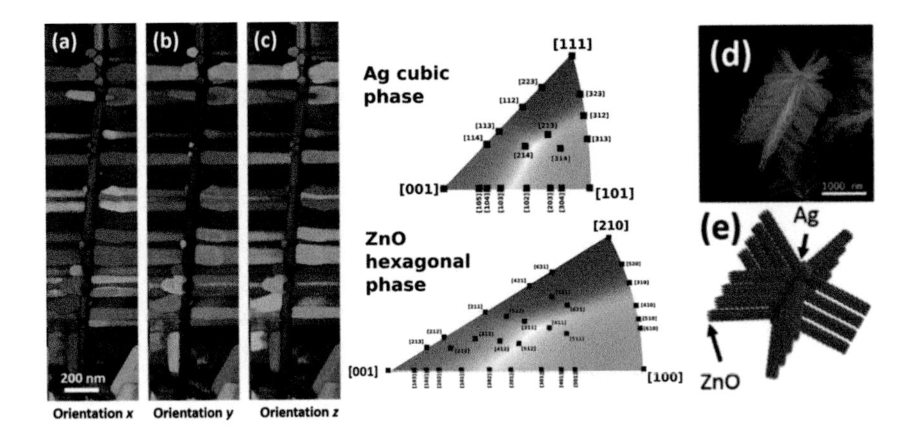

FIGURE 4.10 Crystalline orientation maps using PED ACOM-TEM: orientations (a) *x*, (b) *y*, and (c) *z*. (d) SEM micrograph of the ZnO/Ag assembling and (e) graphical representation of the epitaxial growing.

are oriented along the [001] direction on the (110) planes of the silver. The analysis of the crystalline orientation has been performed by using PED ACOM (Rauch Véron 2014). The combination of PED+ASTAR gives the full crystallographic information of the hybrid assemble. The electron DPs are collected with an external CCD camera connected to the screen viewing of the microscope. Figure 4.10 shows the crystal orientation map; Figure 4.10a–c corresponds to the orientation map in the *x*,*y*,*z*-directions, respectively. Figure 4.10d and e shows a scanning EM (SEM) image and a schematic representation of the assembled hierarchical nanostructure ZnO/Ag. The color chart extracted from the pole figures is shown in Figure 4.10d and e. The color code indicates the orientation near to the [110] zone axis (green color) for the orientation in *x* in silver, which correspond with the parallel direction of the electrons path within the column of the TEM. ZnO nanorods show distinct orientations because the hexagonal rods are rotated in perpendicular directions to the silver nanowires and along their [001] directions.

A detailed analysis of the crystallographic assembling is shown in Figure 4.11a, where an image of a ZnO nanorod was recorded using HRTEM. Figure 4.11b depicts the crystal orientation map of the Ag/ZnO interface; the yellow region corresponds to the ZnO nanorod. The HRTEM image shows that the growth direction corresponds to the planes (002) and that the ZnO nanorod is oriented along the [110] zones axis, as shown in the fast Fourier transform (FFT) inset in Figure 4.11a. The crystallographic relationship at interface is detailed in Figure 4.11b, where three regions of the lateral side of the silver nanowire are projected and show the colors green, red, and blue. To confirm the orientation of the pentagonal cross-sectional area of silver the crystal orientation map of a decahedral particle is added to the image as well as the index map code referred to the fcc area oriented along the fivefold symmetry and its orientation with respect to the ZnO nanorod. The decahedral nanoparticle has five regions, two of them with blue colors oriented along the [111] direction, whereas

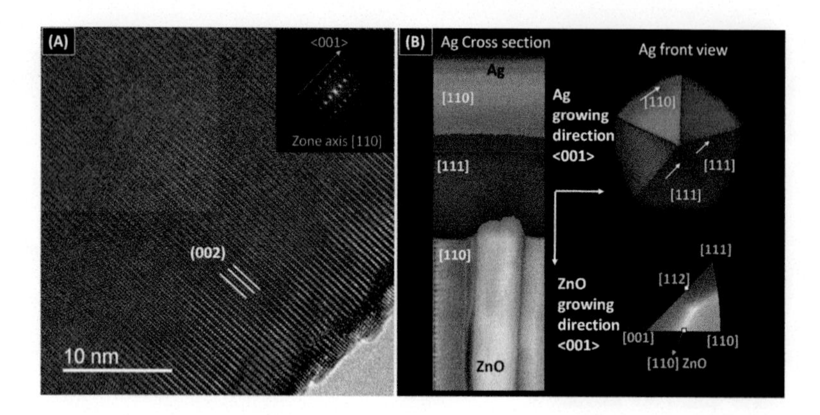

FIGURE 4.11 (a) HRTEM image of ZnO nanorods. The inset shows the FFT, in which the growth direction of the planes (002) along the direction h001i of ZnO nanorods is indicated. (b) Crystal orientation maps show the most probable orientation of a particular phase for Both ZnO nanorods and Ag nanowires, depicted with the color code.

Reprinted with permission from (Sanchez et al. 2016), copyright (2016), AIP.

the green region corresponds to the [110] direction. Similar reports have demonstrated the fivefold zone axis symmetry measured on pentagonal nanoparticles by PED. Furthermore, after the automated indexation process, the planes (002) of the silver nanowire are oriented with the [110] direction of the ZnO nanorod, as shown in Figure 4.11b.

4.3.3 Growth, Crystal Orientation, and Grain Misorientation in Thin Films

A straightforward application of the PED ACOM-TEM technique is the characterization of thin films in which the distortion effects caused by multiple scatterings are reduced and phase contrast dominates the image. Mechanical, functional, and kinetic properties of films mainly depend of crystalline structure parameters in polycrystal systems such as the grain boundary structure determination, size, and texture refinement (Lu 2010). Herein, the evolution of continuous gold thin films, from the formation stage to the structural changes during growth and postannealing treatments, was systematically analyzed. Evolution of structural properties is discussed based on experimental results from the evaluation of grain size, texture, orientation, and misorientation distribution through an ACOM method, and the consequences of recrystallization and grain kinetic processes, primarily grain rotation, and coupled grain boundary migration. Figure 4.12 shows the evolution during the formation and growth of continuous gold films (from 1 to 30 nm in thickness) prepared by thermal evaporation on a NaCl substrate. When gold atoms reached the NaCl surface, they were randomly diffused and formed clusters of adatoms (Yang et al. 1979). Figure 4.12a shows the evolution of a gold thin film growth over a vacuum-cleaved NaCl crystal heated at 200°C. During the first couple of seconds of the deposition process the formation of small decahedral and icosahedral multitwinned particles can be

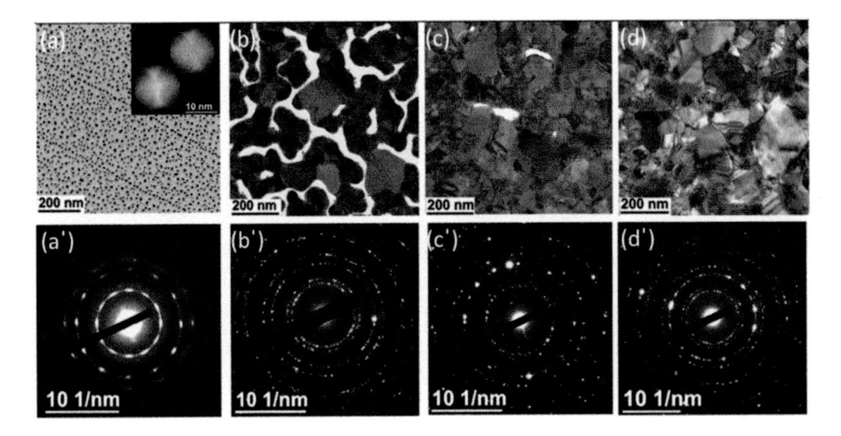

FIGURE 4.12 Gold thin film of different thicknesses indicating stages in the formation of the continuous film. (a) 1 nm (inset is an HRTEM image of a single particle), (b) 15 nm, (c) 25nm, and (d) 30 nm with their corresponding DPs (a′–d′).

Reprinted with permission from (Parajuli et al. 2018), copyright (2019), Wiley.

appreciated (Figure 4.12a, inset), as expected for the evaporation conditions (Ino 1969). The continuous feed of adatoms promotes the formation of small nuclei, which act as seeds to their subsequent growth and cohesion with others until bigger and more stable particles are formed. The structure and shape of gold particles strongly depends on the substrate temperature in which the metal is deposited. The cohesion process continues until the individual gold islands start covering the entire available empty space. The presence of empty gaps, during the island formation process shown in Figure 4.12b, clearly demonstrates the Volmer-Weber growth mode (Venables, Spiller and Hanbucken 1984). This mode is clearly distinguishable from the other two primary modes by which thin films grow epitaxy. Islands continue growing until polycrystalline films are obtained (Figure 4.12c and d). The DP observed in Figure 4.12a presents a strong signal at 4.28 nm^{-1} followed by a dim 4.9 nm^{-1} signal, which jumps to another maximum at 6.94nm^{-1}. These values, which match the Au (111) (200) (220) planes, and the mosaic structure (of two single crystals signals rotated $30 \pm 6°$) highlight the preferential formation of gold nanoparticles oriented over the <101> direction at the beginning of thc polycrystalline thin film formation.

To investigate the grain distribution into gold film samples, a statistical texture analysis was performed by using the ASTAR software package for an as-grown gold film (Figure 4.13a–d) and for an annealed gold film (Figure 4.13a′–d′). Figure 4.13a shows the crystallographic orientation mappings with the grain boundary trace of a 30-nm-thick as-grown film. Each color represents a crystallographic orientation viewed from a particular axis zone. *X*- and *Y*-axes rely on the paper plane, whereas the *Z*-axis was assigned to be parallel to the electron beam in which the experimental DPs were acquired. The color chart code shown inset in Figure 4.13a′ represents a section of the stereographic projection of the different planes along [100] used for orientation maps. With the above example, it is easy to appreciate that Figure 4.13a revealed a predominant [111] texture for the as-grown films, as well as the predominant

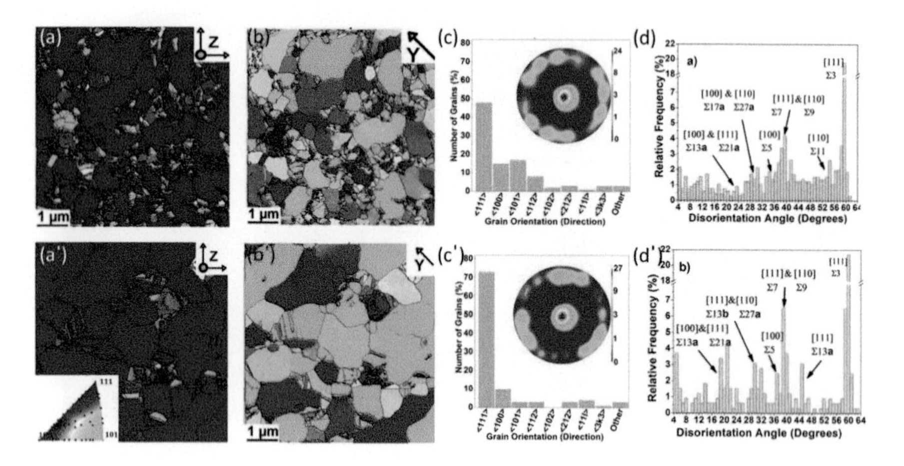

FIGURE 4.13 Combined map of grain boundaries and orientation of an as-grown Au thin film taken along the (a) Z-direction and (b) Y-direction. (c) Grain orientation histogram. This represents relative frequencies of the grains orientated in specific directions. Inset is the pole from Figure 4.13. taken along <111>. (d) Disorientation distribution in as grown film. (a′-d′) Same parameters as shown in (a–d) after annealing has been performed. [Adapted from (Parajuli et al. 2018).]

[101] and [112] orientations when they are viewed along X- and Y-directions, respectively. The plot in Figure 4.13c summarizes the grain orientation of the film from the Z-direction. Thus, the [111] orientation is clearly the most favored, followed by the [101], [001], and [112]. This result is not surprising as it matches the increasing order in energy levels for fcc planes (Vitos et al. 1998). The [111] pole in Figure 4.13 shown as an inset gives information about the orientation of the crystallites in a minimum (blue) to maximum (red) scale. High index orientations were listed as families of direction: [11l], [3k3] and some grains with mixed or no specific direction were labeled as "Other" in the histogram. Numerous studies using TEM and XRD on texture analysis of gold thin films have reported a dominant [111] texture without any additional information on other orientations (Häupl, Lang and Wissmann 1986; Vitos et al. 1998). In this sense, XRD provides overall texture information of a film exposed to the beam, but is not able to give information of single grains. In contrast, TEM diffraction allows the analysis of the orientation of single grains. However, analysis of many grains by obtaining separate DPs of each grain and their subsequent analysis constitutes a long and heavy process. Therefore, the modern ACOM technique allows the analysis of many individual grains with an atomic spatial resolution and provides detailed information about the orientation of each grain.

After annealing of gold film (Figure 4.13a′–d′), the [111] orientation remains as predominant with grains enlarged in size and an increase in the content to more than 70%, respectively, with the almost 50% observed for the as-grown gold film. The other orientations had a significant probability of appearance, and the trend for the minimization of the surface energy face was clearly observed. Grain boundaries in the as-grown film were low coincidence site lattice (CSL), twin, and random high-angle boundaries. Nevertheless, after annealing, random high-angle grain

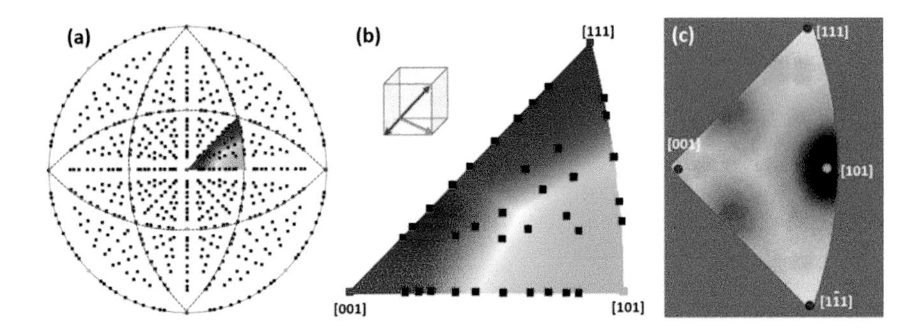

FIGURE 4.14 (a) Cubic inverse pole figure with a symmetrically equivalent region and its color code. (b) Color code chart for the interpretation of crystal orientation maps and the orientation cube. (c) Index matching and cross-correlation map; dark region represents strong matching.

boundaries transformed to low CSL and twin boundaries (TBs). The annealing effect is inferred as the consequence of recrystallization and the kinetic processes of the grain boundary migration tending to a lower energy configuration.

To a better appreciation, Figure 4.14 shows a color chart code (online only) with a triangle formed by the directions [001], [101], and [111] located in the inverse pole diagram (Figure 4.14a). The chart color in Figure 4.14b corresponds to the inverse pole of Figure 4.4. (IPF), which is related to the orientation of the crystal. Therefore, every pixel found in the orientation map viewed along the *Z*-direction (beam direction) whose orientation is close to [001] will appear in the red color and [101] and [111] orientations will be green and blue , respectively, as indicated in Figure 4.14c.

For example, in a similar Au thin film shown in this section, we have postprocessed ASTAR-PED map data acquired in by TEM using the MTEX toolbox in MATLAB (Bachmann, Hielscher and Schaeben 2010). The latter is a freeware toolbox that is extensively used to code and analyze EBSD (Ferreres et al. 2020) datasets. Figure 4.15a shows the inverse pole Figure 4.4 (IPF) map of a gold thin film

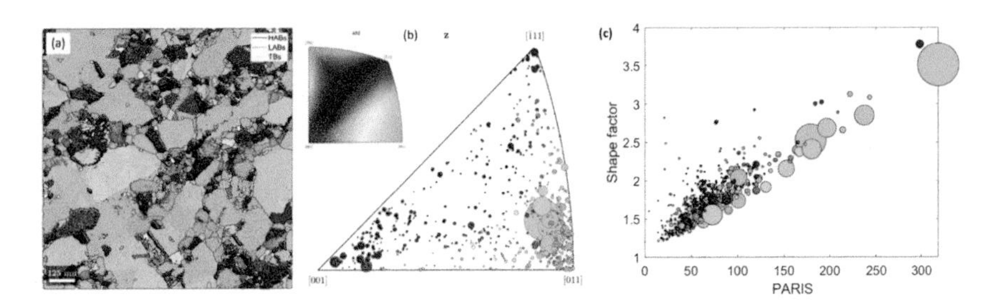

FIGURE 4.15 (a) crystal orientation map of a polycrystalline film taken along the *Z*-direction. (b) IPFz map that shows the proper rotations via stereographic projection and indicating the population of the grains oriented in the polycrystalline film. (c) Plot of shape factor versus the paris of all grains. In (b) and (c) the diameter of the markers is scaled in proportion to the pixel-based grain area.

plotted along the macroscopic z-direction or out of plane from the page surface. To enable ASTAR-PED map data analysis with established EBSD-based analytical techniques, the map data were projected to the fundamental region to describe proper rotations only. Consequently, the color key and legend in Figure 4.15a denote the proper crystal symmetry for an fcc system using Laue or enantiomorphic symmetry groups. Because of the 3 nm step size employed in this map, subgrain structures are defined by a minimum of 20 pixels and are bounded by misorientations $(\theta) \geq 2°$. Consequently, low-angle boundaries (LABs) and high-angle boundaries (HABs) comprise misorientations between $2° \leq \theta \leq 15°$ and $15° \leq \theta \leq 57.5°$, respectively. First-order TBs are defined as $\Sigma 3 = 60°/\langle 111 \rangle$. The maximum tolerance $(\Delta\theta)$ of the misorientation angle from the exact axis-angle twin relationship was identified by the Palumbo-Aust criterion (i.e. $\Delta\theta \leq 15° \ \Sigma^{-5/6}$) such that $\Sigma 3$ boundaries returned a tolerance limit of 6° (Palumbo and Aust 1990). Thus, $\Sigma 3$ boundaries comprise misorientations between $57.5° \leq \theta \leq 62.5°$.

Figure 4.4.15b is a stereographic projection along the macroscopic z-direction that plots the crystallographic axes of each grain by utilizing the color key of Figure 4.15a. Each grain is associated with a specific point in the stereogram and a specific color in the color key; both of which denote its crystal axis pointing out-of-plane from the specimen surface. For each grain orientation, the diameter of its circular marker scales in proportion to its pixel-based grain area. The results show that the grains oriented along the $\langle 110 \rangle$ direction are the most numerous and largest followed by fewer and smaller grains oriented along the $\langle 100 \rangle$ direction and the fewest grains oriented along the $\langle 111 \rangle$ direction.

Figure 4.15c is a plot describing the shape factor versus the percentile average relative indented surface (PARIS) (Panozzo and Hurlimann 1983)) of the grains in Figure 4.15a. Each grain is associated with its specific color in the IPF color key and the diameter of its circular maker scales in proportion to its pixel-based grain area. The shape factor is the ratio between the pixel-based grain perimeter and the equivalent perimeter of its best-fit ellipse. The PARIS factor originates from the Earth Sciences and is a relative measure of the difference between the pixel-based perimeter of a grain and the perimeter/outline of the convex hull/envelope describing its shape, quantifying the lobateness or convexity/concavity of a shape (Hidas et al. 2017). The main advantage of the PARIS factor is that it provides a measure of grain boundary curvature irrespective of grain aspect ratio (Thieme et al. 2018). Thus, fully convex shapes have a PARIS factor of 0. Because this factor is insensitive to the overall grain shape, rounded or angular grains may also possess the equivalent PARIS factors. Plotting the shape and PARIS factors together shows an approximately linear correlation between the two. Grains with small areas tend to have lower shape and PARIS factors whereas grains with larger areas show the opposite trend.

4.4 ELECTRON PAIR DISTRIBUTION FUNCTION APPLIED TO PROTECTED METALLIC CLUSTERS

The previous sections have dealt with standard Bragg diffraction for identification of local crystalline materials. In this section, a refinement of electron pair distribution function (ePDF) structure matching is achieved to study decahedral and truncated-octahedral ligand-protected Au nanoparticles. The traditional crystallographic approach for structure determination results are insufficient for (1) non-crystalline samples; (2) disordered materials; and (3) nanostructures with a clear local structure, but a long-range order over the nanometer regime (poorly defined Bragg diffraction). For these cases, a different approach widely used in x-ray powder diffraction can be implemented for the case of electron diffraction (Hauptman 1986). This analysis method, ePDF, can provide structural information not only in crystalline materials but also in short- and medium-range ordered materials, such as amorphous and nanostructured materials. For the processed DPs it is possible to obtain insights about the atomic structure and volume of the core region, atomic local environment, and the degree of the internal disorder, which affect the overall the degree of crystallinity of materials and nanostructures (Hovmöller, Zou and Weirich 2002; Proffen et al. 2003). Conventional PDF data is extracted form XRD data by synchrotron beamlines. As an example, the advanced light source (ALS) in Berkeley National Lab (Berkeley, California) possesses over 40 different beamlines with different energy ranges from 0.01 to 43 keV (Hexemer et al. 2010). Refinement and fit analyses are compared with the expected theoretical structure to determine the most plausible crystalline structure. The quality of the indexing result is given by a fit-agreement *residual* factor, R_w, resultant from a least-squares model. In the case of gold or other noble metal nanoparticles, an acceptable residual factor is between 0.2 and 0.3 (Petkov et al. 2005; Page et al. 2011; Prasai et al. 2015; Wu et al. 2015; Jensen et al. 2016; Banerjee et al. 2018). As an example, Kumara et al. (2014a) synchrotron x-ray data of $Au_{\sim940}(SR)_{\sim60}$ (mass ~200 kDa, diameter ~3.3 nm) nanoparticles demonstrate an R_w = 44% when compared with a truncated octahedron (TOh) model; whereas their work on $Au_{\sim500}(SR)_{\sim120}$ (~120 kDa) (Kumara et al. 2014b) shows a closer fit of R_w = 37% to a TOh (Figure 4.16a) and a decahedral model with R_w = 23% (Figure 4.16b). A few approaches have been

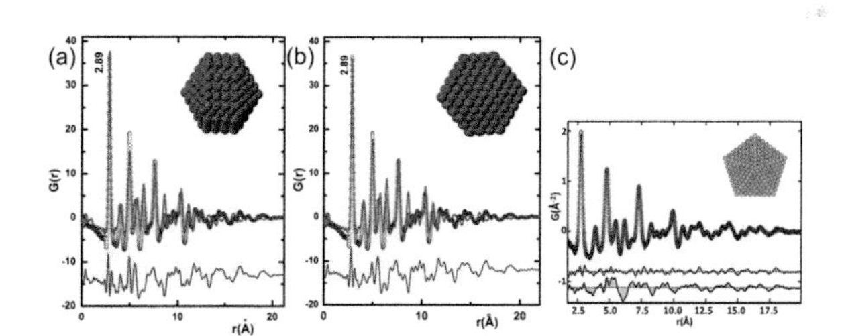

FIGURE 4.16 High-energy x-ray pair distribution function (PDF) of different nanostructures. (a) PDF fitting of Au405 truncated octahedral with the $Au_{\sim500}(SR)_{\sim120}$ cluster (R_w = 37%). (B) Same cluster but under an Au389 Marks decahedral model (R_w = 23%). (C) PDF fitting of a 3.6-nm palladium decahedron model from the experimental Pd nanoparticles. Lower curve differential derived from the use of decahedral (middle line) and spherically attenuated fcc crystal models (bottom line) with R_w = 25.3 and 12.1%, respectively.

reported to improve the fitting result. In the case of allotropy or polymorphism "cluster-mining" uses several structure models instead of a single one, with highly restricted refinement parameters, to obtain the best matching available. In that way they can analyze over 398 different decahedral structures to obtain residual values going from 25.3 to 12.1% as observed in Figure 4.16c (Banerjee et al. 2020). Herein, we propose another approach, using PED to obtain quasi-kinematical electron DP. This allows the application of ePDF measurements under PED-TEM as an alternative to study long-order or disordered nanomaterials in a local range. This approach allows the study of samples using quantities smaller than those required for x-ray synchrotron analysis.

In this section of the chapter, highly monodisperse and stable smaller aqueous gold nanoparticles (core diameter ~4.5 nm) were characterized by the ePDF method. Samples for TEM analyses were prepared on lacey carbon-coated copper grids in a drop-casting (5-μL) fashion and dried in ambient conditions before analysis (Figure 4.17a). The lipoic acid–capped gold nanoparticles (Au-LA nanoparticles) are prepared by the method described by Hoque et al. (2019a,b).

TEM images and SAED patterns were obtained on a JEOL 2010F microscope operating at 200 kV. Figure 4.17b shows a standard SAED DP acquired in a JEOL 2010F microscope operating at 200 kV. The use of a 16-bit CMOS camera ensures a collection of a wide grayscale depth. The electron radiation damage exerted into the lipoic acid–protected ~4.5-nm gold nanoparticles is reduced using a highly saturated CL to reach dose rates of ~8 $\overline{e}/\text{Å}^2\text{s}$. The electron DPs, from larger fields of view (from 250,000 to 640,000 nm^2) are obtained with a PED unit (manufactured by NanoMEGAS) working with 8–25 mrad at 100 Hz.

There are several software packages available to process electron diffraction data to obtain the total structure function, $S(Q)$, or the reduced structure function, $F(Q)$, as well as the reduced PDF, $G(r)$, and normalized PDF, $g(r)$ (Jensen et al. 2016; Tran, Svensson and Tai 2017). In this example, SUePDF and eRDF are being explored to obtain PDF data from electron diffraction (Lábár 2008, 2009; Lábár et al. 2012; Shanmugam et al. 2017; Tran, Svensson and Tai 2017) These programs compute the appropriate electron scattering factors to normalize the intensity and deliver a model for the atomic scattering background to be compared with the experimental data.

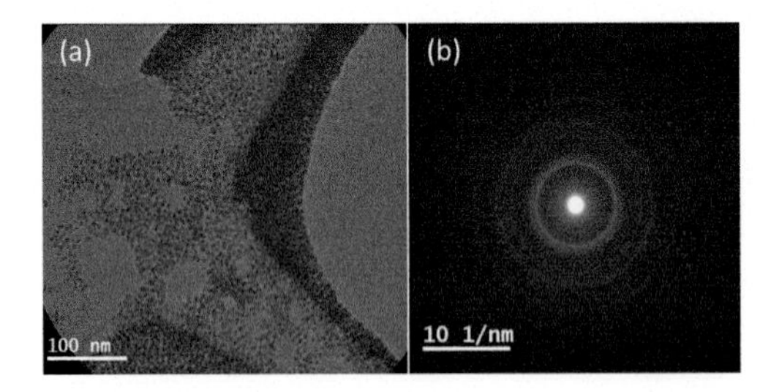

FIGURE 4.17 (a) Electron micrograph of region of interest for electron diffraction. (b) Electron DP of lipoic acid–protected ~4.5-nm gold nanoparticles.

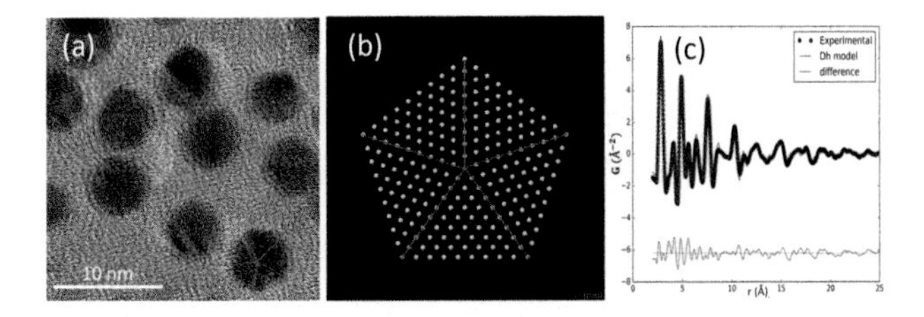

FIGURE 4.18 (a) HRTEM image of 4.5-nm lipoic acid–capped gold nanoparticles (Au-La nanoparticles). (B) Decahedral model structure used for fitting analysis. (c) Best results of PDF fit to the cryogenic PED pattern of a group of Au-La nanoparticles.

To have further confirmation about the quality of the $G(r)$, liquid nitrogen temperatures were used to reduce thermal scattering effects. The Python-based program suite Diffpy-CMI is used to compare the data with the ideal $G(r)$ for Au (Juhás et al. 2015). The residual, R_w, is then calculated with the formula (Banerjee et al. 2018):

$$R_w = \sqrt{\frac{\Sigma_n \left(G_{experimental,n} - G_{theoretical,n}\right)^2}{\Sigma_n \; G^2_{experimental,n}}} \tag{2}$$

Figure 4.18 shows a visualization of the PDF fit analyses of ~4.5 Au nanoparticles applying a bulk fcc model as depicted in Figure 4.18b with its corresponding residual plot (Figure 4.18c). Diffpy-CMI allows variations in the isotropic atomic displacement parameters, generally described as the parameter U_{iso}, permitting the displacement of atoms to optimize the PDF fit. Depending on the resulting values of U_{iso}, the fit's quality can be assessed. These values are reported in Table 4.1, with results comparable with previous reports.

TABLE 4.1
Refined Parameters for PDF Fit Analysis of ~4.5-nm Au Nanoparticles at Room and Cryogenic Temperature.

Conditions/	Room Temperature PED ON/OFF	Liquid Nitrogen Temperature PED ON/OFF			
Parameters	FCC	FCC	Ih	Dh	TOh
U_{iso}(Å^2)	0.031/0.013	0.024/0.018	0.014/0.012	0.019/0.0135	0.023/0.018
Q_{damp}(Å^{-1})	0.088/0.144	0.129/0.105	0.105/0.092	0.103/0.084	0.153/0.101
R_w(%)	29/46	23/39	36/56	22/39	24/39

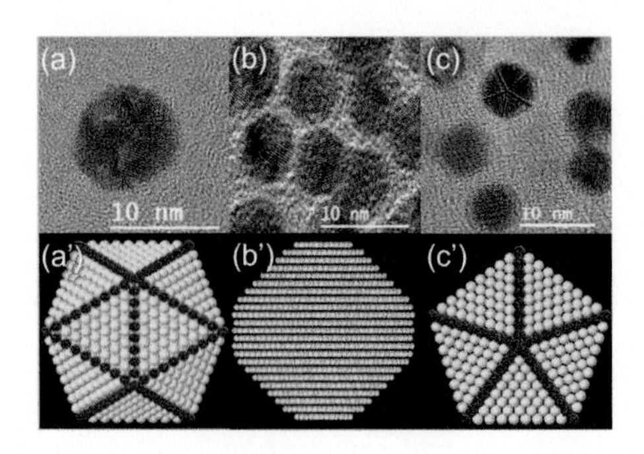

FIGURE 4.19 Structure-models used for ePDF fitting. The three models used (a) icosahedral (Ih), (b) TOh, and (c) decahedral (Dh) are based on the experimental nanoparticles observed under TEM. The corresponding structure models possess (a′) Ih, 2869 Au atoms; (b′) TOh, 2899 Au atoms, and (c′) Dh, 2869 Au atoms.

Figure 4.19 shows the three different structures crossmatched: TOh, icosahedron, and decahedron. The structure models for 4.5-nm Au-LA nanoparticles possess around 3000 atoms. The number of atoms is measured by a structural model that considers the interatomic distances and particle diameter. Au nanoparticles were simulated using "atomic simulation environment" (ASE), a Python module for atomistic simulations (Larsen et al. 2017; Banerjee et al. 2018) and Jmol software for visualization purpose (Herrsez 2006). The TOh model, based on the Wulff construction, uses the surface potentials reported by Zhang, Ma and Xu (2004).

Figure 4.20 compares the three nanoparticle models at different settings. The low temperature condition (without precession) reduced the R_w from 46 to 39%, whereas the minimum residual was achieved with precession and low-temperature conditions (R_w of 23%).

From the analysis of Table 4.1, the TOh and decahedral model provided the best fit with R_w= 24 and 22%, respectively. This is in contrast with the icosahedron results of 40% for 4.2-nm Au nanoparticles and 36% for 4.5-nm Au nanoparticles. From these results, a mixture of different Au nanoparticle shapes is suggested. The presence of different species of Au nanoparticles has been reported in recent work by Banerjee et al. (2018) using x-ray PDF fit analyses for Pd nanoparticles (3.6 nm), $Au_{144}(SC_6)_{44}$ (~2 nm), Au_{144}(p-MBA)$_{60}$, and $Au_{144}(SR)_{60}$ (Yan et al. 2018).

The difference in residual values between the Au clusters reported in the literature and the present colloidal Au nanoparticles relies on clusters being atomically more precise in composition as the ratios (gold and thiol) need to be reached (Jin et al. 2016; Chakraborty and Pradeep 2017). For the nanoparticles, although narrow size distribution synthesis methods have been demonstrated, the nanoparticle composition is still beyond atomic precision (Kumara et al. 2014a, 2018). Notwithstanding, the fit agreements, R_w values below 0.30 obtained with PED and low temperatures, are comparable with the reported high-energy x-ray results (Wu et al. 2015; Jensen et al. 2016; Banerjee et al. 2018).

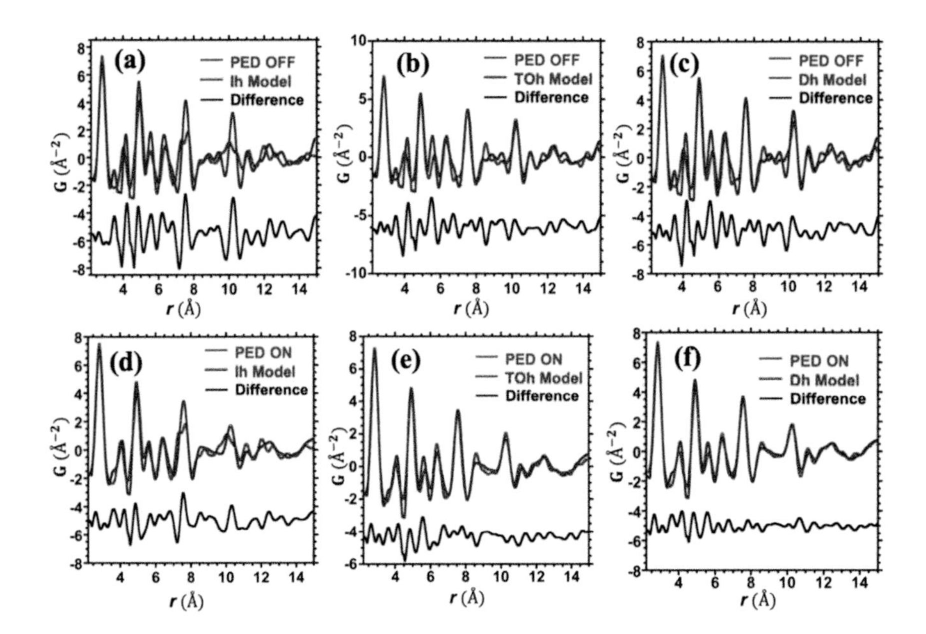

FIGURE 4.20 PDF fit analysis of cryo-electron diffraction results obtained on ~4.5-nm Au nanoparticles, with and without precession, compared with Ih, TOh, and Dh models. Ih: (a) PED OFF, $R_w = 55\%$ and (d) PED ON, $R_w = 36\%$. TOh: (b) PED OFF, $R_w = 39\%$ and (e) PED ON, $R_w = 24\%$ Dh: (c) PED OFF, $R_w = 39\%$ and (f) PED ON, $R_w = 22\%$. The refined parameters of these PDF fit analyses are provided in the Table 4.1.

Reprinted with permission from (Hoque et al. 2019b), copyright (2019), American Chemical Society.

Coordination numbers for the peaks in the PDF were also calculated with and without precession under cryogenic conditions for ~4.5-nm Au nanoparticles using precess diffraction software (Lábár, 2008, 2009; Lábár et al. 2012). In a coordination shell the number of atoms, $N(r)$, is related to the PDF as

$$N(r) = \int_{r_1}^{r_2} 4\pi\rho_0 r^2 g(r)dr \tag{3}$$

$N(r)$ can be retrieved from the experimental $G(r)$ as

$$N(r) = \int_{r_1}^{r_2} [4\pi\rho_0 r^2 + rG(r)]dr \tag{4}$$

For pure gold, any atom will have the first coordination shell composed of 12 neighboring atoms at 2.88 Å. Our first experimental peak was centered at 2.85 Å;

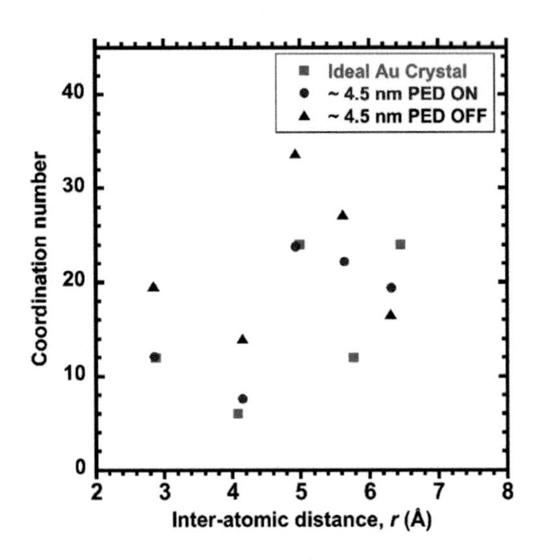

FIGURE 4.21 Coordination number of ~4.5-nm Au nanoparticles with (blue circle) and without (black triangle) PED in comparison with ideal Au crystals (red square).

the integral corresponding to this peak resulted in a coordination number of 12.1 under precession, whereas the unprecessed case resulted in 19.6 atoms, as shown in Figure 4.21.

Combination of PED and low temperatures reduces the least-square residuum R_w, associated with the percentage of residual error, with respect to electron DPs collected without precession at room temperature. The improvement of the kinematical enhancement using only PED is about 23%, and an additional 11% of the reduction is achieved at low temperatures. The superior fitting of the experiential data and model structure are also confirmed by a better estimation of the coordination number from ePDF-PED data under cryogenic conditions. It is shown that the carbon film contribution that supports the nanoparticles plays an important role in the determination of the R_w residuum, and for that reason a background subtraction must be applied. Acquisition of the electron DPs using low-dose electrons prevents damage in the ligand-protected gold nanoparticles and gives reproducible and consistent values. Finally, we conclude that the cryogenic and PED measurements are an essential tool that overcomes some of the limitations in the conventional x-ray PDF methods, especially for nanocrystals, and opens up a new door to study the structure of other types of nanostructured materials using TEM rather than the synchrotron x-ray sources.

4.5 CONCLUDING REMARKS

Figure 4.22 summarizes how the large dataset electron DPs are generated. A single electron DP obtained in any of the electron diffraction modalities contains information of the crystalline structure of the samples, symmetry properties, identification of phases, and information of the PDF mainly considered for complex structures. However, the large number of patterns starts when the beam is irradiating the

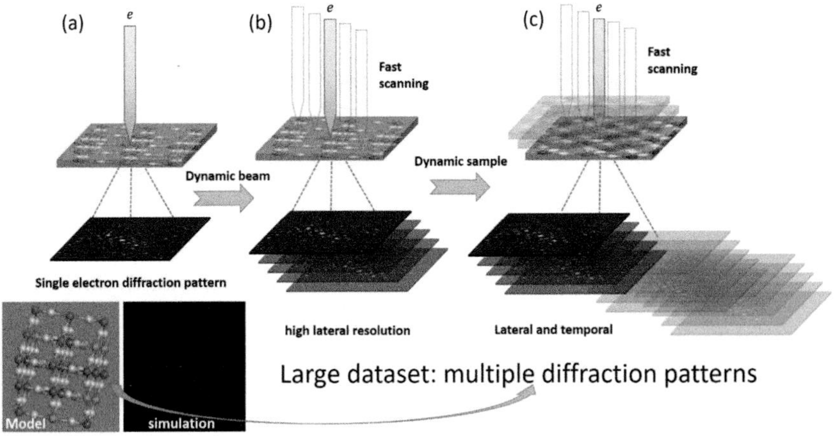

FIGURE 4.22 Schematic sequence of (a) a single electron DP under a static beam, (b) the production of multiple DPs due to the dynamic nanobeam under fast scanning with a sample fix, and (c) fast scanning nanobeam diffraction interacting with a dynamic specimen (sensitive materials).

sample with a sharp probe and is deflected along the field of view line by line. The challenge is in part the frequency in which the electron bean can be moved and the capability of registering the patterns in a fast detector with enough quality to register the maximum number of reflections. For nanostructures, it is more critical because the signal-to-noise ratio increases and the analysis become more complicated. The structural modeling of the specimen supports the analysis of crystals. For sensitive materials, such as proteins, zeolites, and ligand-protected clusters, the irradiation can be static using ultralow electron doses and decreasing the accelerating voltage. However, in many cases these procedures do not reduce the radiation damage and the fast scanning reduces the interaction time with the specimen. The evolution of the sample is due to the momentum lost by the beam transferred to the crystal (Marks and Zhang 1992). The momentum transferred produces a change in the orientation of the crystals, and the number of DPs is even greater. The examples of the methodologies herein give a broad view of the spectrum of possibilities to study materials at the resolution limit by means of new developments in instrumentation, including the controlled dosage of electrons and the DE detection, all pushing the limit in spatial and temporal resolution.

ACKNOWLEDGMENTS

The authors gratefully acknowledge the financial support provided by U.S. Department of Defense W911NF-18-1-0439 and #72489 as well as to the VPREDKE at UTSA. Authors also gratefully thank to CONACYT for the support under the postdoctoral fellowship program 382259 and 740562. We acknowledge also to the Secretary of Education, Science, Technology, and Innovation (SECTEI) from Mexico City under the postdoctoral fellowship program SECTEI/300/2019.

REFERENCES

Allé, P., Wenger, E., Dahaoui, S., Schaniel, D. and Lecomte, C., 2016. Comparison of CCD, CMOS and hybrid pixel x-ray detectors: detection principle and data quality. *Physica Scripta*, *91*(6): 063001.

Bachmann, F., Hielscher, R. and Schaeben, H., 2010. Texture analysis with MTEX–free and open source software toolbox. *Solid State Phenomena* (160): 63–68.

Bahena, D., Bhattarai, N., Santiago, U., Tlahuice, A., Ponce, A., Bach, S.B., Yoon, B., Whetten, R.L., Landman, U., and Jose-Yacaman, M., 2013. STEM electron diffraction and high-resolution images used in the determination of the crystal structure of the Au144 (SR) 60 cluster. *The Journal of Physical Chemistry Letters*, *4*(6): 975–981.

Baletto, F., and Ferrando, R., 2005. Structural properties of nanoclusters: energetic, thermodynamic, and kinetic effects. *Reviews of Modern Physics*, *77*(1): 371.

Banerjee, S., Liu, C.H., Jensen, K.M., Juhás, P., Lee, J.D., Tofanelli, M., Ackerson, C.J., Murray, C.B. and Billinge, S.J., 2020. Cluster-mining: an approach for determining core structures of metallic nanoparticles from atomic pair distribution function data. *Acta Crystallographica Section A: Foundations and Advances*, *76*(1): 24–31.

Banerjee, S., Liu, C.H., Lee, J.D., Kovyakh, A., Grasmik, V., Prymak, O., Koenigsmann, C., Liu, H., Wang, L., Abeykoon, A.M. and Wong, S.S., 2018. Improved models for metallic nanoparticle cores from atomic pair distribution function (PDF) analysis. *The Journal of Physical Chemistry C*, *122*(51): 29498–29506.

Barnard, A.S., 2014. Clarifying stability, probability and population in nanoparticle ensembles. *Nanoscale*, *6*(17): 9983–9990.

Betal, S., Saha, A.K., Ortega, E., Dutta, M., Ramasubramanian, A.K., Bhalla, A.S., and Guo, R., 2018. Core-shell magnetoelectric nanorobot–A remotely controlled probe for targeted cell manipulation. *Scientific Reports*, *8*(1): 1–9.

Bruma, A., Santiago, U., Alducin, D., Plascencia Villa, G., Whetten, R.L., Ponce, A., Mariscal, M., and José-Yacamán, M., 2016. Structure determination of superatom metallic clusters using rapid scanning electron diffraction. *The Journal of Physical Chemistry C*, *120*(3): 1902–1908.

Carbó-Argibay, E., Rodríguez-González, B., Pastoriza-Santos, I., Pérez-Juste, J., and Liz-Marzán, L.M., 2010. Growth of pentatwinned gold nanorods into truncated decahedra. *Nanoscale*, *2*(11): 2377–2383.

Casallas-Moreno, Y.L., Cardona, D., Ortega, E., Hernández-Gutiérrez, C.A., Gallardo-Hernández, S., Hernández-Hernández, L.A., Gómez-Pozos, H., Ponce, A., Contreras-Puente, G., and López-López, M., 2018. High cubic phase purity and growth mechanism of cubic InN thin-films by migration enhanced epitaxy. *Thin Solid Films*, *647*: 64–69.

Chakraborty, I. and Pradeep, T., 2017. Atomically precise clusters of noble metals: emerging link between atoms and nanoparticles. *Chemical Reviews*, *117*(12): 8208–8271.

Champness, P.E., 1987. Convergent beam electron diffraction. *Mineralogical Magazine*, *51*(359): 33–48.

Cowley, J.M., 1996. Electron nanodiffraction: progress and prospects. *Microscopy*, *45*(1): 3–10.

Cowley, J.M., 1999. Electron nanodiffraction. *Microscopy Research and Technique*, *46*(2): 75–97.

Cowley, J.M., Janney, D.E., Gerkin, R.C. and Buseck, P.R., 2000. The structure of ferritin cores determined by electron nanodiffraction. *Journal of Structural Biology*, *131*(3): 210–216.

Darbal, A.D., Ganesh, K.J., Liu, X., Lee, S.B., Ledonne, J., Sun, T., Yao, B., Warren, A.P., Rohrer, G.S., Rollett, A.D. and Ferreira, P.J., 2013. Grain boundary character distribution of nanocrystalline Cu thin films using stereological analysis of transmission electron microscope orientation maps. *Microscopy and Microanalysis*, *19*(1): 111–119.

Deng, Y., Tasan, C.C., Pradeep, K.G., Springer, H., Kostka, A. and Raabe, D., 2015. Design of a twinning-induced plasticity high entropy alloy. *Acta Materialia, 94*: 124–133.

Egerton, R.F., 2013. Control of radiation damage in the TEM. *Ultramicroscopy, 127*: 100–108.

Egerton, R.F., 2019. Radiation damage to organic and inorganic specimens in the TEM. *Micron, 119*: 72–87.

Ferreres, X.R., Casillas, G., Aminorroaya-Yamini, S. and Gazder, A.A., 2020. Multiphase identification in Ni–PbTe contacts by EBSD and aberration-corrected STEM. *Materials & Design, 185*: 108252.

Fish, D.A., Brinicombe, A.M., Pike, E.R. and Walker, J.G., 1995. Blind deconvolution by means of the Richardson–Lucy algorithm. *JOSA A, 12*(1): 58–65.

Ganesh, K.J., Kawasaki, M., Zhou, J.P., and Ferreira, P.J., 2010. D-STEM: a parallel electron diffraction technique applied to nanomaterials. *Microscopy and Microanalysis, 16*(5): 614–621.

GIMP-team GIMP (GNU Image Manipulation Program).

Gu, X. and Mildner, D.F.R., 2016. Ultra-small-angle neutron scattering with azimuthal asymmetry. *Journal of Applied Crystallography, 49*(3): 934–943.

Häupl, K., Lang, M. and Wissmann, P., 1986. X-Ray diffraction investigations on ultra-thin gold films. *Surface and Interface Analysis, 9*(1): 27–30.

Hauptman, H., 1986. The direct methods of X-ray crystallography. *Science, 233*(4760): 178–183.

Hawkes, P.W., 2015. The correction of electron lens aberrations. *Ultramicroscopy, 156*: A1–A64.

Herraez, A., 2006. Biomolecules in the computer: Jmol to the rescue. *Biochemistry and Molecular Biology Education, 34*(4): 255–261.

Hexemer, A., Bras, W., Glossinger, J., Schaible, E., Gann, E., Kirian, R., MacDowell, A., Church, M., Rude, B., and Padmore, H., 2010. A SAXS/WAXS/GISAXS beamline with multilayer monochromator. *Journal of Physics: Conference Series, 247*(1): 012007.

Hidas, K., Tommasi, A., Mainprice, D., Chauve, T., Barou, F., and Montagnat, M., 2017. Microstructural evolution during thermal annealing of ice-Ih. *Journal of Structural Geology, 99*: 31–44.

Hoque, M.M., Mayer, K.M., Ponce, A., Alvarez, M.M. and Whetten, R.L., 2019a. Toward smaller aqueous-phase plasmonic gold nanoparticles: high-stability thiolate-protected~4.5 nm cores. *Langmuir, 35*(32): 10610–10617.

Hoque, M.M., Vergara, S., Das, P.P., Ugarte, D., Santiago, U., Kumara, C., Whetten, R.L., Dass, A. and Ponce, A., 2019b. Structural analysis of ligand-protected smaller metallic nanocrystals by atomic pair distribution function under precession electron diffraction. *The Journal of Physical Chemistry C, 123*(32): 19894–19902.

Hovmöller, S., Zou, X., and Weirich, T.E., 2002. Crystal Structure Determination from EM Images and Electron Diffraction Patterns. In *Advances in Imaging and Electron Physics*, Hawkes, P. W. Merli, P. G. Calestani, G. Vittori-Antisari, M., Eds. Elsevier: Vol. 123, pp. 257–289.

Ino, S., 1969. Stability of multiply-twinned particles. *Science Reports of the Research Institutes, Tohoku University. Ser. A, Physics, Chemistry and Metallurgy, 22*: 128.

Jensen, K.M., Juhas, P., Tofanelli, M.A., Heinecke, C.L., Vaughan, G., Ackerson, C.J., and Billinge, S.J., 2016. Polymorphism in magic-sized Au 144 (SR) 60 clusters. *Nature Communications, 7*(1): 1–8.

Jin, R., Zeng, C., Zhou, M., and Chen, Y., 2016. Atomically precise colloidal metal nanoclusters and nanoparticles: fundamentals and opportunities. *Chemical Reviews, 116*(18): 10346–10413.

Juhás, P., Farrow, C.L., Yang, X., Knox, K.R. and Billinge, S.J., 2015. Complex modeling: a strategy and software program for combining multiple information sources to solve ill posed structure and nanostructure inverse problems. *Acta Crystallographica Section A*, *71*(6): 562–568.

Kobler, A., Kashiwar, A., Hahn, H. and Kübel, C., 2013. Combination of in situ straining and ACOM TEM: a novel method for analysis of plastic deformation of nanocrystalline metals. *Ultramicroscopy*, *128*: 68–81.

Kolb, U., Mugnaioli, E., and Gorelik, T.E., 2011. Automated electron diffraction tomography–a new tool for nano crystal structure analysis. *Crystal Research and Technology*, *46*(6): 542–554.

Kumara, C., Hoque, M.M., Zuo, X., Cullen, D.A., Whetten, R.L. and Dass, A., 2018. Isolation of a 300 kDa, $au_{\sim 1400}$ gold compound, the standard 3.6 nm capstone to a series of plasmonic nanocrystals protected by aliphatic-like thiolates. *The Journal of Physical Chemistry Letters*, *9*(23): 6825–6832.

Kumara, C., Zuo, X., Cullen, D.A. and Dass, A., 2014a. Faradaurate-940: synthesis, mass spectrometry, electron microscopy, high-energy X-ray diffraction, and X-ray scattering study of au~ 940±20 (SR)~ 160±4 nanocrystals. *ACS Nano*, *8*(6): 6431–6439.

Kumara, C., Zuo, X., Ilavsky, J., Chapman, K.W., Cullen, D.A. and Dass, A., 2014b. Super-stable, highly monodisperse plasmonic faradaurate-500 nanocrystals with 500 gold atoms: $au_{\sim 500}$ $(SR)_{\sim 120}$. *Journal of the American Chemical Society*, *136*(20): 7410–7417.

Lábár, J.L., 2005. Consistent indexing of a (set of) single crystal SAED pattern (s) with the ProcessDiffraction program. *Ultramicroscopy*, *103*(3): 237–249.

Lábár, J.L., 2008. Electron diffraction based analysis of phase fractions and texture in nanocrystalline thin films, part I: principles. *Microscopy and Microanalysis*, *14*(4): 287–295.

Lábár, J.L., 2009. Electron diffraction based analysis of phase fractions and texture in nanocrystalline thin films, part II: implementation. *Microscopy and Microanalysis*, *15*(1): 20–29.

Lábár, J.L., Adamik, M., Barna, B.P., Czigány, Z., Fogarassy, Z., Horváth, Z.E., Geszti, O., Misják, F., Morgiel, J., Radnóczi, G. and Sáfrán, G., 2012. Electron diffraction based analysis of phase fractions and texture in nanocrystalline thin films, part III: application examples. *Microscopy and Microanalysis*, *18*(2): 406–420.

Larsen, A.H., Mortensen, J.J., Blomqvist, J., Castelli, I.E., Christensen, R., Dułak, M., Friis, J., Groves, M.N., Hammer, B., Hargus, C. and Hermes, E.D., 2017. The atomic simulation environment—a python library for working with atoms. *Journal of Physics: Condensed Matter*, *29*(27): 273002.

Legorreta-Flores, A., Davila-Tejeda, A., Velásquez-González, O., Ortega, E., Ponce, A., Castillo-Michel, H., Reyes-Grajeda, J.P., Hernández-Rivera, R., Cuéllar-Cruz, M. and Moreno, A., 2018. Calcium carbonate crystal shapes mediated by intramineral proteins from eggshells of ratite birds and crocodiles. Implications to the eggshell's formation of a dinosaur of 70 million years old. *Crystal Growth & Design*, *18*(9): 5663–5673.

Lu, K., 2010. The future of metals. *Science*, *328*(5976): 319–320.

Magnan, P., 2003. Detection of visible photons in CCD and CMOS: a comparative view. *Nuclear Instruments and Methods in Physics Research Section A: Accelerators, Spectrometers, Detectors and Associated Equipment*, *504*(1–3): 199–212.

Maksymov, I.S., 2015. Magneto-plasmonics and resonant interaction of light with dynamic magnetisation in metallic and all-magneto-dielectric nanostructures. *Nanomaterials*, *5*(2): 577–613.

Maksymov, I.S., 2016. Magneto-plasmonic nanoantennas: basics and applications. *Reviews in Physics*, *1*: 36–51.

Marks, L.D., 1984. Surface structure and energetics of multiply twinned particles. *Philosophical Magazine A*, *49*(1): 81–93.

Marks, L.D. and Zhang, J.P., 1992. Is there an electron wind? *Ultramicroscopy*, *41*(4): 419–422.

Matsui, H., and Tabata, H., 2012. Lattice strains and polarized luminescence in homoepitaxial growth of a-plane ZnO. *Applied Physics Letters*, *101*(23).

Method of the Year 2015, 2016. Editorial. *Nature Methods*, 13(1): 1–1.

Moeck, P. and Rouvimov, S., 2010. Precession electron diffraction and its advantages for structural fingerprinting in the transmission electron microscope. *Zeitschrift für Kristallographie-Crystalline Materials*, *225*(2–3): 110–124.

Moeck, P., Rouvimov, S., Rauch, E.F., Véron, M., Kirmse, H., Häusler, I., Neumann, W., Bultreys, D., Maniette, Y. and Nicolopoulos, S., 2011. High spatial resolution semi-automatic crystallite orientation and phase mapping of nanocrystals in transmission electron microscopes. *Crystal Research and Technology*, *46*(6): 589–606.

Ortega, E., Ponce, A., Santiago, U., Alducin, D., Benitez-Lara, A., Plascencia-Villa, G. and José-Yacamán, M., 2017a. Structural damage reduction in protected gold clusters by electron diffraction methods. *Advanced Structural and Chemical Imaging*, *2*(1): 12.

Ortega, E., Reddy, S.M., Betancourt, I., Roughani, S., Stadler, B.J. and Ponce, A., 2017b. Magnetic ordering in 45 nm-diameter multisegmented FeGa/Cu nanowires: single nanowires and arrays. *Journal of Materials Chemistry C*, *5*(30): 7546–7552.

Ortega, E., Santiago, U., Giuliani, J.G., Monton, C. and Ponce, A., 2018. In-situ magnetization/heating electron holography to study the magnetic ordering in arrays of nickel metallic nanowires. *AIP Advances*, *8*(5): 056813.

Ortega, J.E., Santiago, U., Bruna, A., Alducin, D., Villa, G.P., Whetten, R.L., Ponce, A. and Jose-Yacaman, M., 2016. Fast scanning electron diffraction and electron holography as methods to acquire structural information on Au102 (p-MBA) 44 nanoclusters. *Microscopy and Microanalysis*, *22*(S3): 528–529.

Page, K., Hood, T.C., Proffen, T. and Neder, R.B., 2011. Building and refining complete nanoparticle structures with total scattering data. *Journal of Applied Crystallography*, *44*(2): 327–336.

Palumbo, G. and Aust, K.T., 1990. Structure-dependence of intergranular corrosion in high purity nickel. *Acta Metallurgica et Materialia*, *38*(11): 2343–2352.

Panozzo, R. and Hurlimann, H., 1983. A simple method for the quantitative discrimination of convex and convex-concave lines. *Microscopica Acta*, *87*(2): 169–176.

Pantzer, A., Vakahy, A., Eliyahou, Z., Levi, G., Horvitz, D. and Kohn, A., 2014. Dopant mapping in thin FIB prepared silicon samples by off-axis electron holography. *Ultramicroscopy*, *13:* 836–45.

Parajuli, P., Mendoza-Cruz, R., Santiago, U., Ponce, A. and Yacamán, M.J., 2018. The evolution of growth, crystal orientation, and grain boundaries disorientation distribution in gold thin films. *Crystal Research and Technology*, *53*(8): 1800038.

Petkov, V., Peng, Y., Williams, G., Huang, B., Tomalia, D. and Ren, Y., 2005. Structure of gold nanoparticles suspended in water studied by x-ray diffraction and computer simulations. *Physical review B*, *72*(19): 195402.

Prasai, B., Wilson, A.R., Wiley, B.J., Ren, Y. and Petkov, V., 2015. On the road to metallic nanoparticles by rational design: bridging the gap between atomic-level theoretical modeling and reality by total scattering experiments. *Nanoscale*, *7*(42): 17902–17922.

Proffen, T., Billinge, S.J.L., Egami, T. and Louca, D., 2003. Structural analysis of complex materials using the atomic pair distribution function—A practical guide. *Zeitschrift für Kristallographie-Crystalline Materials*, *218*(2): 132–143.

Quintana, C., Cowley, J.M. and Marhic, C., 2004. Electron nanodiffraction and high-resolution electron microscopy studies of the structure and composition of physiological and pathological ferritin. *Journal of Structural Biology*, *147*(2): 166–178.

Rauch, E.F., Portillo, J., Nicolopoulos, S., Bultreys, D., Rouvimov, S. and Moeck, P., 2010. Automated nanocrystal orientation and phase mapping in the transmission electron microscope on the basis of precession electron diffraction. *Zeitschrift für Kristallographie-Crystalline Materials*, *225*(2–3): 103–109.

Rauch, E.F. and Veron, M., 2005. Coupled microstructural observations and local texture measurements with an automated crystallographic orientation mapping tool attached to a TEM. *Materialwissenschaft und Werkstofftechnik: Entwicklung, Fertigung, Prüfung, Eigenschaften und Anwendungen technischer Werkstoffe*, *36*(10): 552–556.

Rauch, E.F. and Véron, M.J.M.C., 2014. Automated crystal orientation and phase mapping in TEM. *Materials Characterization*, *98*: 1–9.

Ringe, E., Van Duyne, R.P. and Marks, L.D., 2013. Kinetic and thermodynamic modified Wulff constructions for twinned nanoparticles. *The Journal of Physical Chemistry C*, *117*(31): 15859–15870.

Ruiz-Zepeda, F., Casallas-Moreno, Y.L., Cantu-Valle, J., Alducin, D., Santiago, U., José-Yacaman, M., López-López, M. and Ponce, A., 2014. Precession electron diffraction-assisted crystal phase mapping of metastable c-GaN films grown on (001) GaAs. *Microscopy Research and Technique*, *77*(12): 980–985.

Sanchez, J.E., González, G., Vera-Reveles, G., Velazquez-Salazar, J.J., Bazan-Diaz, L., Gutiérrez-Hernández, J.M., José-Yacaman, M., Ponce, A. and González, F.J., 2017. Silver/zinc oxide self-assembled nanostructured bolometer. *Infrared Physics & Technology*, *81*: 266–270.

Sanchez, J.E., Santiago, U., Benitez, A., Yacamán, M.J., González, F.J. and Ponce, A., 2016. Structural analysis of the epitaxial interface Ag/ZnO in hierarchical nanoantennas. *Applied Physics Letters*, *109*(15): 153104.

Sang, X., Kulovits, A. and Wiezorek, J., 2013. Comparison of convergent beam electron diffraction methods for simultaneous structure and Debye Waller factor determination. *Ultramicroscopy*, *126*: 48–59.

Santiago, U., Velázquez-Salazar, J.J., Sanchez, J.E., Ruiz-Zepeda, F., Ortega, J.E., Reyes-Gasga, J., Bazán-Díaz, L., Betancourt, I., Rauch, E.F., Veron, M. and Ponce, A., 2016. A stable multiply twinned decahedral gold nanoparticle with a barrel-like shape. *Surface Science*, *644*: 80–85.

Shanmugam, J., Borisenko, K.B., Chou, Y.J. and Kirkland, A.I., 2017. eRDF analyser: an interactive GUI for electron reduced density function analysis. *SoftwareX*, *6*: 185–192.

Shigesato, G. and Rauch, E.F., 2007. Dislocation structure misorientations measured with an automated electron diffraction pattern indexing tool. *Materials Science and Engineering: A*, *462*(1–2): 402–406.

Stadelmann, P.A., 1987. EMS-a software package for electron diffraction analysis and HREM image simulation in materials science. *Ultramicroscopy*, *21*(2): 131–145.

Steuwer, A., Santisteban, J.R., Turski, M., Withers, P.J. and Buslaps, T., 2004. High-resolution strain mapping in bulk samples using full-profile analysis of energy-dispersive synchrotron X-ray diffraction data. *Journal of Applied Crystallography*, *37*(6): 883–889.

Thieme, M., Demouchy, S., Mainprice, D., Barou, F. and Cordier, P., 2018. Stress evolution and associated microstructure during transient creep of olivine at 1000–1200 C. *Physics of the Earth and Planetary Interiors*, *278*: 34–46.

Tlahuice-Flores, A., Santiago, U., Bahena, D., Vinogradova, E., Conroy, C.V., Ahuja, T., Bach, S.B., Ponce, A., Wang, G., José-Yacamán, M. and Whetten, R.L., 2013. Structure of the thiolated Au130 cluster. *The Journal of Physical Chemistry A*, *117*(40): 10470–10476.

Tran, D.T., Svensson, G. and Tai, C.W., 2017. SUePDF: a program to obtain quantitative pair distribution functions from electron diffraction data. *Journal of Applied Crystallography*, *50*(1): 304–312.

Tschumperlé, D., and Fourey, S., 2016. GMIC: GREYCs Magic for Image Computing: A full-featured open-source framework for image processing. https://gmic.eu.

TVIPS-GmbH TemCam XF-Series. https://www.tvips.com/camera-systems/temcam-xf-series/ (accessed 11/04/2019).

Venables, J.A., Spiller, G.D.T., and Hanbucken, M., 1984. Nucleation and growth of thin films. *Reports on Progress in Physics*, *47*(4): 399–459.

Viladot, D., Véron, M., Gemmi, M., Peiró, F., Portillo, J., Estradé, S., Mendoza, J., Llorca-Isern, N. and Nicolopoulos, S., 2013. Orientation and phase mapping in the transmission electron microscope using precession-assisted diffraction spot recognition: state-of-the-art results. *Journal of Microscopy*, *252*(1): 23–34.

Vincent, R. and Midgley, P.A., 1994. Double conical beam-rocking system for measurement of integrated electron diffraction intensities. *Ultramicroscopy*, *53*(3): 271–282.

Vitos, L., Ruban, A.V., Skriver, H.L. and Kollar, J., 1998. The surface energy of metals. *Surface Science*, *411*(1–2): 186–202.

Wu, J., Shan, S., Petkov, V., Prasai, B., Cronk, H., Joseph, P., Luo, J. and Zhong, C.J., 2015. Composition–structure–activity relationships for palladium-alloyed nanocatalysts in oxygen reduction reaction: an ex-situ/in-situ high energy X-ray diffraction study. *ACS Catalysis*, *5*(9): 5317–5327.

Yan, N., Xia, N., Liao, L., Zhu, M., Jin, F., Jin, R. and Wu, Z., 2018. Unraveling the long-pursued Au144 structure by x-ray crystallography. *Science Advances*, *4*(10): eaat7259.

Yang, C.Y., Heinemann, K., Yacaman, M.J. and Poppa, H., 1979. A structural analysis of small vapor-deposited "multiply twinned" gold particles. *Thin Solid Films*, *58*(1): 163–168.

Zankel, A., Kraus, B., Pölt, P., Schaffer, M. and Ingolic, E., 2009. Ultramicrotomy in the ESEM, a versatile method for materials and life sciences. *Journal of Microscopy*, *233*(1): 140–148.

Zhang, J.-M., Ma, F. and Xu, K.-W., 2004. Calculation of the surface energy of FCC metals with modified embedded-atom method. *Applied Surface Science*, 229(1–4): 34–42.

Zhou, Z., Liu, X., Liu, Q. and Liu, L., 2008. Evaluation of the potential cytotoxicity of metals associated with implanted biomaterials (I). *Preparative Biochemistry and Biotechnology*, *39*(1): 81–91.

Zou, X., Hovmöller, S., and Oleynikov, P., 2011. *Electron Crystallography: Electron Microscopy and Electron Diffraction*. Oxford University Press, Vol. 16.

Index

A

ACOM-TEM, 119–120, 125–128
AICrystallographer, 47
Amorphous layers, 18–19
Analog-to-digital converter, 3
Annular dark-field detector, 2
  inner- and outer-angle, 12
Artificial intelligence, 41
Artificial neural network, 43
ASTAR, 115, 127, 129, 131–132

B

Beam, 4, 6, 8, 9–10, 12, 16, 18–20, 23, 26–27, 47, 55, 57–58, 70–74, 112–115, 119–126, 129, 138–139
  broadening, 6, 17, 18
  current, 11, 20
  damage, 18, 27
  deflection, 52
  heating, 18
  sensitive, 29, 71
  stopper, 28

C

Camera length, 8, 12–13
Carbon background, 18
Cation ordering, 91–92
CBED, 72, 113
Chemical composition, 69
Chemical state, 73
Clathrate, 78–80
Cold field emission current fluctuations, 20
Complex ordering, 95–100
Condenser lens, 9, 110, 114, 115
Confocal, 8
Cross talk, 17–18
Crystal structure, 70–72, 75, 79, 81, 85, 87, 88
CuInSe, 80–84
Custom-scanning design, 30–31

D

Data requirements, 57–58
Data speed, 61
Data-centric requirements, 61–63
Deep convolutional neural network, 44
Deep learning, 42, 44–47, 62
Deep neural networks, 43
DeepDiffraction, 47
Defects, 72
Detector, 2–15, 17, 20, 22, 23, 28–30, 42, 57, 59, 61, 74, 139
Detector response linearity, 11
Diffraction, 47, 50, 55, 70–73, 75, 78–80, 83, 85–86, 88, 111
Digital super-resolution, 27–28
Dimensionality reduction in STEM, 49–57
Dropped current, 11
Dropped gain, 12

E

EDX, 3–4, 17, 25–27, 31, 70–71, 73–74, 76, 78–79, 83, 91, 102
EELS, 3, 10, 12, 26–28, 31, 73–74, 78, 79–81, 88, 91, 93, 94–96, 98–100, 103–104
Efficiency response, 10
EFTEM, 73
Electron dose, 19
Emission decay, 20
Environmental noise, 20–24
Experimental intensities, 16–19

F

Flux distribution, 13–15

G

GitHub, 47
Graphical processing unit, 43, 61, 64

H

HAADF-STEM, 72–93, 95–104
Hollandite, 84–90
Hybrid analysis approaches, 4–8

I

Intensity, 2–6, 11, 14, 16–22, 24, 50, 52, 55, 77–78, 83, 87–88, 91–93, 96–97, 113–115, 122, 125, 134,

J

JEMS, 120

K

Kikuchi, 29

L

LAADF, 4, 18, 29
Layered ordering, 97–101

M

Machine learning for electron microscopy, 41–63
Magnetic structure, 73
Manganoferrite, 75–78
Material genome initiative, 42
MEMS, 126
Moore's law, 41
Morphology, 72, 81, 83
Multi-frame applications, 24
Multiframe imaging, 28–31

N

NanoMEGAS, 125, 134
Nanoparticles, 80–83
Neutron diffraction, 80, 112
Normalization approach, 13–15

O

Oxygen-deficient perovskite, 95–101

P

Pair distribution function, 116, 132–138
Peak positions, 16–21
PED-TEM, 134
Perovskite, 90–103
Pixel density, 27–28
Pixel size, 19
pnDetector, 31
Precession
  precession electron diffraction (PED), 72, 115, 123, 124, 125, 136, 137, 138
Proteins, 117–119, 139
py4DSTEM, 47
pyCroscopy, 47
pyUSID, 47

Q

Quantitative ADF, 2, 27–30
Quasi-equilibrium, 73
Quintuple, 95–100

R

Rare-earth cations, 78–80
Recording, ADF detector scans, 8

S

SAED, 70, 112–113, 119, 123, 134
Sample tilt, 16–17
Scan patterning, 30–31
Scanning distortion in the STEM, 20–23
Simulation reference, 4
SNR, 24
Sources of errors, 20
Spectroscopy, 4, 24, 28, 43, 70, 73, 74, 71
Statistical decomposition, 4
StatSTEM, 7
STEM, 2–10, 12, 14, 16–17, 19–22, 24, 27, 29–31, 42–45, 48–49, 51–53, 58–59, 72–104, 114, 120, 125
STEM imaging, 24, 25
Strain contrast, 18
Supervised learning in STEM, 44–48
Swung-beam detector scanning, 8

T

TEM, 112–121, 125, 127, 130–131, 134, 136, 138
TVIPS, 120

U

Ultrafast process, 73
Unsupervised learning in STEM, 49–57

V

Vacancy, 86, 88, 96, 98, 100, 103

W

Water removal, 118
Watershed, 6
Window, 50

X

X-ray crystallography, 117
X-ray data, 133
X-ray diffraction, 4, 17, 67, 70, 128
X-ray scattering, 75, 78
X-ray spectroscopy, 3, 70
X-ray tomography, 58

Y

Yield, 3, 7, 13, 16, 20, 27, 29, 88

Z

Z-contrast, 18, 44, 74